ÍNDICE

Nota del autor

This material contains enough theoretical development and proposed exercises with solutions, in order to promote our students' autonomy. Moreover, for almost each introduced concept, some exercises are solved step by step.

To prepare this material, I have used the following bibliography:

Bibliography	- Fundamentals of Mathematics (Authors: J. Van Dyke, J. Rogers, H. Adams; Ed.: BROOKS/COLE, CENGAGE Learning),
	- Spanish text books:
	    - Matemáticas 4º ESO (Autores: J. Cólera y otros; Ed.: Anaya).
	    - Notes and work sheets (in English and Spanish): Jose Alonso Manzanera, Francisco J. Glez Ortiz.
	    - English Dictionary (Oxford Edition).

This materials have been reviewed in order to fulfil new Spanish **LOMCE** regulations.

Para el profesor: <u>Evaluación por estándares:</u> He preparado un cuadro en el que relaciono cada estándar evaluable de la LOMCE con los ejercicios de este libro que lo trabajan, y que puede ser muy útil a la hora de diseñar las pruebas de evaluación. Si estás interesado en que te lo envíe, ponte en contacto conmigo en el mail de más abajo.

En mi web (www.cuadernodepitagoras.com) he subido las presentaciones que uso con mis alumnos en clase. Si en algo crees que te puedo ser de utilidad, si encuentras erratas, o si deseas hacerme alguna sugerencia, te agradeceré que te pongas en contacto conmigo en el mail javiersanchezpi@gmail.com.

Soy doctor en Química y licenciado en Ciencias Ambientales y soy profesor de matemáticas en un instituto en Murcia (España).

Espero te sea de utilidad y que contactes conmigo para lo que precises.

Un saludo.

Javier Sánchez Pina.

$$\sqrt{2} \qquad \sqrt[3]{2} \qquad \pi$$

$$\sqrt{1+\sqrt{2}}$$

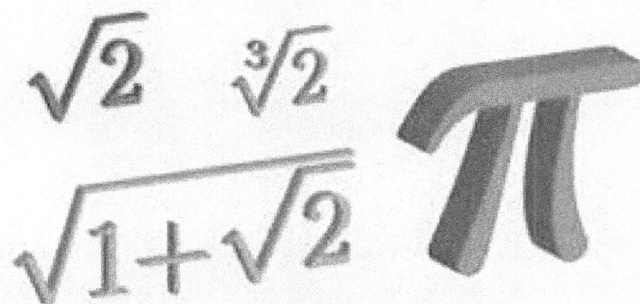

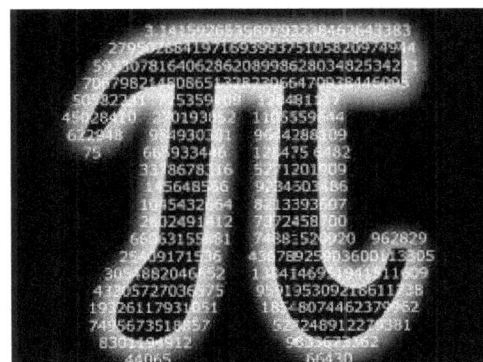

Unit 1.- Real numbers

1. Sets of numbers

Quantities are classified into different sets or type of numbers.

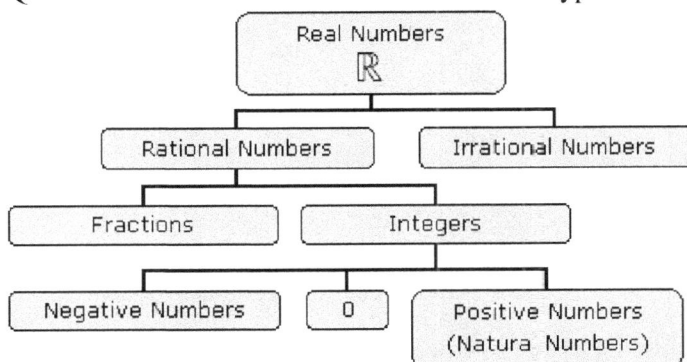

Rational numbers (Q): They are those numbers that can be written as a fraction. A fraction is the indicated quotient of two integer numbers. There are several types of rational numbers:

- **Integer numbers (Z):** $Z = \{...., -3, -2, -1, 0, 1, 2, 3, ...\}$.

The set of the integer number is can be classified, as well, into two sets, called **natural numbers (N)** = $\{0, 1, 2, 3, ...\}$ and their opposites $(Z^-) = \{-1, -2, -3, ...\}$, called as **negative integers**.

All these numbers are rational because we can write them as a fraction.

- **Non integer rational numbers:** They are all those fractions whose quotient is NOT a whole number. For example, $\dfrac{1}{2}$, $\dfrac{4}{3}$ or $\dfrac{7}{10}$.

When calculating their decimal expressions, we always obtain a decimal number (an exact or a recurring decimal number):

– **Exact or terminating** decimal: It finishes, so you can write down all its digits.

For example: $\dfrac{7}{4} = 7 \div 4 = 1.75$

– **Recurring or repeating** decimal: It does not finish, it goes on forever, but <u>some of the digits are repeated over and over again</u>. For example:

$$5.6666...... = 5.\overline{6} \qquad\qquad 14.353535........ = 14.\overline{35} \qquad\qquad 87.42222...... = 87.4\overline{2}$$

We can distinguish:

- **Pure recurring decimals**: Recurring decimals in which period starts just after the decimal point.

For example: $5.6666...... = 5.\overline{6}$ $\qquad\qquad 14.353535........ = 14.\overline{35}$

- **Mixed recurring decimals**: Recurring decimals in which period does not start just after the decimal point. For example: $87.42222...... = 87.4\overline{2}$

Rational expression of decimal numbers

In the inverse way, the decimal expression of a rational number can be converted into a fraction:

All decimal numbers can be expressed as a fraction.

In the following example we show the process to find out these fractions.

- **Exact decimal:** Express as a fraction $x = 2.34$

<u>Solution:</u> We must build a fraction whose

- Numerator: Decimal number without decimal point.
- Denominator: A "1" followed by as many "0" as decimal digits decimal number has.

In our case: $2.34 = \dfrac{234}{100}$ If possible, we must simplify this fraction to give it in its simplest form.

$$2.34 = \dfrac{224}{100} = \dfrac{112}{50} = \boxed{\dfrac{56}{25}}$$

- Recurring decimal number: Express as a fraction $x = 14.\overline{35} = 14.3535355\ldots\ldots$

Solution: We must apply the following process, consisting on multiplying it by 1, 10, 100, 1000, …., till finding two decimals having _identical decimal parts_, so that they can be subtracted, to cancel their decimal parts and converting them into integers.

$$100 \cdot x = 1435.353535\ldots$$
$$10 \cdot x = 143.535353\ldots$$
$$x = 14.353535\ldots$$

As you can see, x and 100·x have the same decimal part. So, we subtract them:

$$100 \cdot x = 1435.353535\ldots$$
$$\cancel{10 \cdot x = 143.535353\ldots}$$
$$x = 14.353535\ldots$$

$$99 \cdot x = 1421 \implies x = \frac{1421}{99}$$

Example: Express as a fraction: a) $8.\overline{5}$ b) $25.4\overline{52}$

Solution:

a)

$$10 \cdot x = 85.555\ldots$$
$$x = 8.5555\ldots$$

As you can see, both decimals have the same decimal part. So, we subtract them:

$$10 \cdot x = 85.555\ldots$$
$$x = 8.5555\ldots$$

$$9 \cdot x = 77 \implies x = \frac{77}{9}$$

b)

$$1000 \cdot x = 25452.5252\ldots$$
$$100 \cdot x = 2545.25252\ldots$$
$$10 \cdot x = 254.525252\ldots$$
$$x = 25.4525252\ldots$$

As you can see, 1000·x and 10·x have the same decimal part. So, we subtract them:

$$1000 \cdot x = 25452.5252\ldots$$
$$\cancel{100 \cdot x = 2545.25252\ldots}$$
$$\cancel{10 \cdot x = 254.525252\ldots}$$
$$x = 25.4525252\ldots$$

$$990 \cdot x = 25198 \implies x = \frac{25198}{990}$$

– Irrational numbers (I): They are those numbers that CANNOT be written as a fraction. Irrationals are those decimal numbers having an infinite quantity decimal digits, non being periodically repeated. Some examples of irrational numbers:
$3,1010010001\ldots.$; $0,3737737773\ldots..$;
$0,123456789101112\ldots\ldots$

$$\pi = 3.14159265\ldots$$
$$9502884197169\ldots$$
$$07816406286208\ldots$$
$$0982148086513\ldots$$
$$9821720553\ldots$$

Some other more famous irrational numbers are:

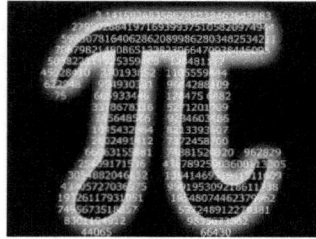

Л = 3,141592654...... is obtained when dividing the length of a circumference by its diameter.

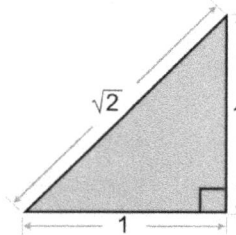

$\sqrt{2}$ = 1,414213562....... is obtained when calculating the diagonal of a square whose side is 1 and $\sqrt[3]{2}$ = 1,25992105...... is obtained when calculating the edge of a cube whose volume is 2.

e = 2.718281828…….. is the base of the natural logarithms.

They are also irrational numbers those roots whose result is not an integer or a recurring decimal, as well as the results of adding, subtracting, multiplying or dividing a rational and an irrational number.

Exercises

1. Write as a fraction when possible:

$-2;\ \ 1,\overline{7};\ \ \sqrt{3};\ \ 4,\overline{2};\ \ -3,75;\ \ 3\pi;\ \ -2\sqrt{5}$

Which of the numbers below are irrational numbers?

2. Classify the following numbers into rational or irrational numbers and arrange them for lower to higher:

$\dfrac{\sqrt{3}}{2};\ \ 0,8\overline{7};\ \ -\sqrt{4};\ \ -\dfrac{7}{3};\ \ \dfrac{1}{\sqrt{2}};\ \ 2\pi$

3. Locate the following numbers into their correct place:

$$3,42; \quad \frac{5}{6}; \quad -\frac{3}{4}; \quad \sqrt{81}; \quad \sqrt{5}; \quad -1; \quad \frac{\pi}{4}; \quad 1,4555...$$

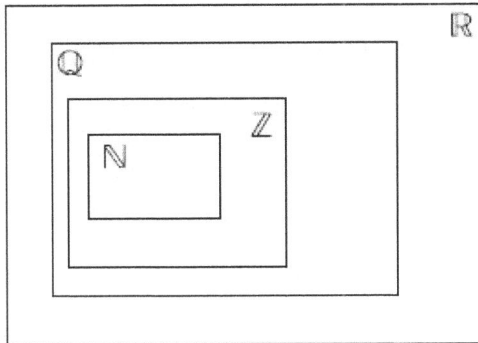

4. Express as a fraction:

a) 23.45 b) 5.101 c) $24.\overline{3}$ d) $0.\overline{678}$ e) $32.7\overline{6}$ f) $1.\overline{76}$

g) $26.\overline{7}$ h) $0.2\overline{3}$ i) $2.7\overline{5}$ j) $34.6\overline{41}$ k) $2.09\overline{5}$ l) $5.1\overline{204}$

m) $1.0\overline{4}$ n) $7.\overline{123}$ ñ) $8.17\overline{4}$ o) $2.5\overline{19}$

5. Find the reason why period 9 is never considered, this is, $0.\overline{9}$ or $0.0\overline{9}$ do not make sense.

6. Work out, directly and writing first as fractions, and check you obtain the same results:

a) $0.\overline{3}+0.\overline{6}$ b) $0.\overline{3}-0.\overline{15}$ c) $3.\overline{41}+2.37\overline{8}$ d) $0.\overline{4}\cdot0.1$ e) $3.\overline{1}+2.0\overline{3}$

f) $0.\overline{3}+0.1\overline{6}$ g) $4\cdot2.\overline{5}$ h) $4.\overline{89}-3.\overline{78}$ i) $8-2.\overline{7}$ j) $4.5\cdot0.0\overline{2}+0.\overline{4}$

2. Intervals in the Real Line

Intervals are segments and semi-straight lines into the Real Line.

There are different types of intervals:

The open interval **(a,b)** includes the real numbers x between a and b, but a and b are NOT included. It is expressed as a < x < b.

The closed interval **[a,b]** includes the real numbers x between a and b, being a and b included. It is expressed as $a \leq x \leq b$.

The open semi-straight line **(a, ∞)** includes all the real numbers being higher than a, where a is not included. It is expressed as $a < x$;

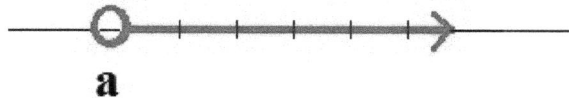

In the same way, the interval **[a,b)** is expressed as $a \leq x < b$, being a included but not b;

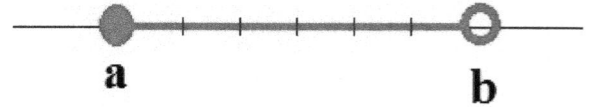

while the interval **(a,b]** is expressed as $a < x \leq b$, being b included but not a.

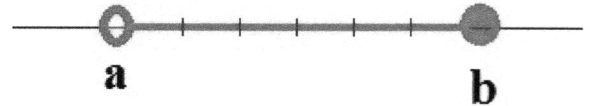

while the closed semi-straight line **[a, ∞)** includes all the real numbers being higher than a, being a included. It is expressed as $a \leq x$.

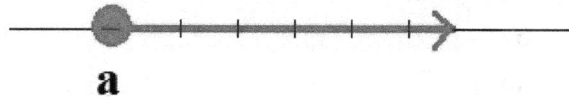

In the same way, semi-straight line **(-∞, b)** includes all the real numbers being lower than b, where b is not included. It is expressed as $x < b$;

while the semi-straight line **[-∞, b]** includes all the real numbers being lower than b, being b included. It is expressed as $x \leq b$.

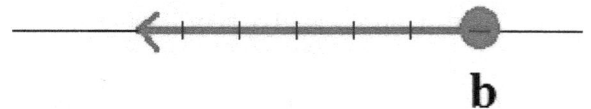

7. On the Real Line, represent these intervals:

A = [−2, 4]   B = (1, 6)   C = [−7, −3)   D = (0, 5]

E = (−∞, 1]   F = (−1, +∞)

8. Write as an interval and represent on the Real Line those numbers that fulfil the given inequalities:

a) −3 ≤ x ≤ 2   b) −1 < x < 5   c) 0 < x ≤ 7   d) x > −5

9. Express as an inequality and as an interval the sets of numbers represented as follows:

a)

b)

c)

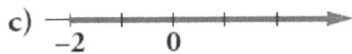

d)

10. Write, in all the possible ways, the following intervals:

a) {x / (−2) ≤ x < 3}   b) (−∞, (−2)]   c) Numbers higher than (−1)   d)

11. Represent on an only Real Line the following semilines:

A = (−∞, 3] and B = [−1, +∞)

Which are the numbers that are included at the same time in A and B ($A \cap B$)? Express it as an interval.

12. Represent on the Real Line:

a) (−∞, −3) ∪ (1, +∞)   b) (−∞, 0] ∪ [2, +∞)

---

3. Operations with rational numbers

As you have widely studied operations with fractions, in the next chart you are offered a brief summary of them, as well as some examples.

3.1.- Addition and subtraction of fractions

- If they have the same denominator, we will leave the denominator and add or subtract the numerators.

$$\dfrac{a}{b} + \dfrac{c}{b} = \dfrac{a+c}{b}$$ Example: $\dfrac{2}{5} + \dfrac{1}{5} = \boxed{\dfrac{3}{5}}$

- If they have different denominators, first of all, we will transform them onto new fractions having all of them the same denominator, which will be the LCM of the original denominators.

Example: $\dfrac{1}{4} - \dfrac{2}{3} + \dfrac{7}{5} \Rightarrow LCM(4,3,5) = 60 \Rightarrow \dfrac{15}{60} - \dfrac{40}{60} + \dfrac{84}{60} = \boxed{\dfrac{59}{60}}$

3.2.- Product of fractions

We must multiply, separately, numerator by numerator and denominator by denominator.

$$\frac{a}{b} \cdot \frac{c}{d} = \frac{a \cdot c}{b \cdot d}$$

Example: Calculate $\dfrac{6}{8} \cdot \dfrac{10}{15}$

Solution:

Simplifying at the end) $\quad \dfrac{6}{8} \cdot \dfrac{10}{15} = \dfrac{6 \cdot 10}{8 \cdot 15} = \dfrac{60}{120} = \dfrac{6}{12} = \dfrac{6 \div 2}{12 \div 2} = \dfrac{3 \div 3}{6 \div 3} = \boxed{\dfrac{1}{2}}$

Simplifying before) $\dfrac{6}{8} \cdot \dfrac{10}{15} = \dfrac{6 \cdot 10}{8 \cdot 15} = \dfrac{(2 \cdot 3) \cdot (2 \cdot 5)}{(2 \cdot 2 \cdot 2) \cdot (3 \cdot 5)} = \dfrac{2 \cdot 3 \cdot 2 \cdot 5}{2 \cdot 2 \cdot 2 \cdot 3 \cdot 5} = \boxed{\dfrac{1}{2}}$

3.3.- Inverse of a fraction

We obtain the inverse of a fraction, changing the places of numerator and denominator.

Inverse of $\dfrac{a}{b}$ is $\boxed{\dfrac{b}{a}}$

3.4.- Division of fractions

We multiply the first fraction by the inverse of the second one.

$$\frac{a}{b} : \frac{c}{d} = \frac{a}{b} \cdot \frac{d}{c} = \frac{a \cdot d}{b \cdot c}$$

Example: Calculate $\dfrac{6}{8} : \dfrac{15}{10}$

Solution:

Simplifying at the end) $\quad \dfrac{6}{8} : \dfrac{15}{10} = \dfrac{6}{8} \cdot \dfrac{10}{15} = \dfrac{6 \cdot 10}{8 \cdot 15} = \dfrac{60}{120} = \dfrac{6}{12} = \dfrac{6 \div 2}{12 \div 2} = \dfrac{3 \div 3}{6 \div 3} = \boxed{\dfrac{1}{2}}$

Simplifying before) $\dfrac{6}{8} : \dfrac{15}{10} = \dfrac{6}{8} \cdot \dfrac{10}{15} = \dfrac{6 \cdot 10}{8 \cdot 15} = \dfrac{(2 \cdot 3) \cdot (2 \cdot 5)}{(2 \cdot 2 \cdot 2) \cdot (3 \cdot 5)} = \dfrac{2 \cdot 3 \cdot 2 \cdot 5}{2 \cdot 2 \cdot 2 \cdot 3 \cdot 5} = \boxed{\dfrac{1}{2}}$

3.5.- Mixed operations with fractions

They will be done following this order:
- 1st) Products and divisions (from left to right).
- 2nd) Additions and subtractions.
- If there are parentheses or brackets, we will solve first into them.

13. Arrange the following numbers, from lower to higher:

a) $\dfrac{1}{2} \quad \dfrac{3}{4} \quad \dfrac{5}{6}$

b) $\dfrac{1}{2} \quad \dfrac{3}{5} \quad \dfrac{7}{15}$

c) $\dfrac{1}{5} \quad \dfrac{3}{4} \quad \dfrac{-2}{7} \quad \dfrac{9}{8} \quad \dfrac{6}{5} \quad \dfrac{5}{6}$

14. Work out:

a) $\dfrac{4}{5} + \dfrac{1}{6} - \dfrac{2}{15}$

b) $2 + \dfrac{5}{6} + \dfrac{1}{3}$

c) $\dfrac{-3}{10} + \dfrac{5}{6} + \dfrac{4}{3}$

d) $\dfrac{-5}{12} + \dfrac{2}{9} + \dfrac{-7}{15}$

e) $1 - \dfrac{-2}{3} + 3$

f) $\left(\dfrac{3}{4} - \dfrac{1}{5} \right) - \left(\dfrac{-7}{2} + \dfrac{5}{2} \right)$

15. Work out:

a) $\dfrac{2}{7} \cdot \dfrac{3}{5}$

b) $\dfrac{2}{3} \cdot \dfrac{5}{7} \cdot \dfrac{6}{9}$

c) $\dfrac{2}{7} : \dfrac{3}{5}$

d) $\left(\dfrac{2}{7} : \dfrac{4}{5} \right) \cdot \dfrac{4}{6}$

e) $\dfrac{15}{2} : (-5)$

f) $\left(\dfrac{3}{4} \cdot \dfrac{1}{5} \right) + \left(\dfrac{-7}{2} : \dfrac{5}{2} \right)$

16. A student says: «During a day, I spend 1/3 of time sleeping and 3/8 at school. Classes take the 2/3 of the time I spend at school». What fraction of my day do classes take?

17. Work out by two methods, directly and extracting a common factor:

a) $\dfrac{2}{3} \cdot \dfrac{1}{5} + \dfrac{2}{3} \cdot \dfrac{5}{6}$

b) $\dfrac{3}{4} \cdot \dfrac{5}{6} + \dfrac{5}{6} \cdot \dfrac{1}{9}$

c) $\dfrac{-3}{11} \cdot \dfrac{1}{4} - \dfrac{6}{5} \cdot \dfrac{-3}{11}$

18. Work out the following expressions:

a) $3 - 4 \left[\dfrac{1}{3} - \dfrac{1}{2} \left(\dfrac{1}{4} - \dfrac{1}{5} \right) + 3 : \left(\dfrac{1}{3} : \dfrac{1}{2} \right) \right]$

b) $(3 - 4) \left[\left(\dfrac{1}{3} - \dfrac{1}{2} \right) \left(\dfrac{1}{4} - \dfrac{1}{5} \right) + \left(3 : \dfrac{1}{3} \right) : \dfrac{1}{2} \right]$

c) $\left(\dfrac{1}{3} + \dfrac{1}{2} \right) \left(\dfrac{1}{2} - \dfrac{1}{4} \right) + 5 - 3 \left(4 : \dfrac{3}{5} + 1 \right)$

d) $\left[\dfrac{1}{3} + \dfrac{1}{2} \left(\dfrac{1}{2} - \dfrac{1}{4} \right) + 5 \right] - 3 \left[4 : \left(\dfrac{3}{5} + 1 \right) \right]$

e) $\left[\dfrac{3}{8} \left(\dfrac{5}{3} - \dfrac{1}{2} \right) - \dfrac{4}{11} \left(\dfrac{3}{4} - \dfrac{1}{5} \right) \right] : \left[\dfrac{5}{9} - \left(\dfrac{-3}{4} + \dfrac{1}{2} \right) + \dfrac{10}{3} \left(\dfrac{1}{2} - \dfrac{3}{5} \right) \right]$

f) $\left(\dfrac{3}{5} : \dfrac{2}{3} - \dfrac{4}{5} \cdot \dfrac{4}{3} + \dfrac{1}{3} - \dfrac{3}{4} : \dfrac{3}{7} \right) \left(\dfrac{2}{3} + \dfrac{-7}{2} - \dfrac{5}{6} + \dfrac{1}{4} \right) : \left(\dfrac{-4}{3} + \dfrac{2}{3} - \dfrac{1}{6} \right)$

4. Scientific notation

When we observe Nature out of quotidian life, for example going into microscopic world or going out, to other planets and galaxies, we are finding quantities that are not comfortable to use. Sometimes because they are very little (size of a cell = 0.0000000018 cm) and sometimes because they are too large (light speed = 300 000 000 m/s).

These quantities, containing long chains of zeros, to the right of to the left of significant digits, make them uncomfortable to use.

In order to make them more comfortable to handle, it is used *scientific notation*, which consists on transforming numbers with long chains, into shorter expressions. In the following examples, we are showing how scientific notation works.

Parts of a number in scientific notation, $N = a.bcc... \times 10^k$:
- An integer part, with an only digit (Not zero).
- A decimal part.
- A power with base 10 and integer exponent. (k is an integer, positive or negative).

Example: Express 322400000000000 in scientific notation.

Step 1: Write significant digits, with a decimal point after the first of them. **Step 2:** Count how many positions the decimal point has been moved. This is going to be the exponent.	3.224 322400000000000.0000....... 3,224000000000000000000 We have moved decimal point 14 positions. So, $3.224 \cdot 10^{14}$
Step 3: <u>Sign of the exponent.</u> There is a very easy rule: - If initial quantity was higher than new one with decimal point, exponent is positive; - If initial quantity was lower than new one with decimal point, the exponent will be negative.	In this case, as 322400000000000 > 3,224, exponent stays positive: $3.224 \cdot 10^{14}$

This transformation can also be done by using the definition of power:

Example: Express in scientific notation these quantities: a) 529000   b) 0.00025
<u>Solution:</u>

    a) $529000 = 5.29 \cdot 100000 = 5.29 \cdot 10^5$ b) $0.00025 = \dfrac{2.5}{100000} = 2.5 \cdot \dfrac{1}{100000} = 2.5 \cdot 10^{-5}$

Exercises	**19.** Write in scientific notation:
	a) 657 b) 0.00058 c) 1258000 d) 0.0021 e) 321000 f) 0.000012 g) 0.0012 h) 7800 i) 9757000 j) 0.00023

Product and quotient in scientific notation

To *multiply* or *divide* two quantities given in scientific notation, we multiply or divide, separately, numeric parts and powers of 10. You will have to use the laws of powers.

Example: Calculate: a) $2.7 \cdot 10^{-5} \cdot 3.2 \cdot 10^{8}$ b) $\dfrac{4.62 \cdot 10^{7}}{2 \cdot 10^{-4}}$ c) $8.31 \cdot 10^{-5} \cdot 5.2 \cdot 10^{9}$

Solution:

a) $2.7 \cdot 10^{-5} \cdot 3.2 \cdot 10^{8} = \begin{vmatrix} 2.7 \cdot 3.2 = 8.64 \\ 10^{-5} \cdot 10^{8} = 10^{-5+8} = 10^{3} \end{vmatrix} = 8.64 \cdot 10^{3}.$

b) $\dfrac{4.62 \cdot 10^{7}}{2 \cdot 10^{-4}} = \begin{vmatrix} \dfrac{4.62}{2} = 2.31 \\ 10^{7} : 10^{-4} = 10^{7-(-4)} = 10^{11} \end{vmatrix} = 2.31 \cdot 10^{11}.$

c) $8.31 \cdot 10^{-5} \cdot 5.2 \cdot 10^{9} = \begin{vmatrix} 8.31 \cdot 5.2 = 43.212 \\ 10^{-5} \cdot 10^{9} = 10^{-5+9} = 10^{4} \end{vmatrix} = 43.212 \cdot 10^{4}$, but in scientific notation, we can

write only one significant digit before decimal point, so, we will write $4.3212 \cdot 10 \cdot 10^{4} = 4.3212 \cdot 10^{5}.$

Exer cises	**20.** Work out: a) $4,1 \cdot 10^{3} \times 2 \cdot 10^{5}$ b) $2,8 \cdot 10^{-4} : 2 \cdot 10^{3}$ c) $3,2 \cdot 10^{5} : 2 \cdot 10^{-2}$

Addition and subtraction of numbers in scientific notation

When *adding* or *subtracting* numbers in scientific notation, we need all the powers of 10 to have the **same exponents**. In this case, we will add or subtract the numerical part, leaving the same power of 10.
* If they do not have the same exponent, previously, we must reduce to *common exponent*.

Example: Calculate: a) $3.75 \cdot 10^{8} + 2.11 \cdot 10^{8}$ b) $2.72 \cdot 10^{5} - 3.1 \cdot 10^{3}$

Solution:
a) $3.75 \cdot 10^{8} + 2.11 \cdot 10^{8} = (3.75 + 2.11) \cdot 10^{8} = 5.86 \cdot 10^{8}.$
b) $2.72 \cdot 10^{5} - 3.1 \cdot 10^{3} =$ Notice they do not have the same exponents. So, we must reduce to common exponent. And we can choose:

$2.72 \cdot 10^{5} - 3.1 \cdot 10^{3} = \begin{vmatrix} 2.72 \cdot 10^{5} = 2.72 \cdot 10^{2} \cdot 10^{3} = 272 \cdot 10^{3} \end{vmatrix} = 272 \cdot 10^{3} - 3.1 \cdot 10^{3} =$

$= (272 - 3.1) \cdot 10^{3} = 268.9 \cdot 10^{3} = 2.689 \cdot 10^{5}.$

$2.72 \cdot 10^{5} - 3.1 \cdot 10^{3} = \begin{vmatrix} 3.1 \cdot 10^{3} = 3.1 \cdot 10^{-2} \cdot 10^{5} = 0.031 \cdot 10^{5} \end{vmatrix} = 2.72 \cdot 10^{5} - 0.031 \cdot 10^{5} =$

$= (2.72 - 0.031) \cdot 10^{5} = 2.689 \cdot 10^{5}.$

21. Express as scientific notation:

a) $32 \cdot 10^5$ b) $75 \cdot 10^{-4}$ c) $843 \cdot 10^7$ d) $458 \cdot 10^{-7}$ e) $0,03 \cdot 10^6$ f) $0,0025 \cdot 10^{-5}$

22. Calculate, without using your paper and pencil:

a) $(1,5 \cdot 10^7) \cdot (2 \cdot 10^5)$ b) $(3 \cdot 10^6):(2 \cdot 10^{11})$ c) $(4 \cdot 10^{-7}):(2 \cdot 10^{-12})$ d) $\sqrt{4 \cdot 10^8}$

23. Calculate, expressing the result as scientific notation, and check your answer by using your calculator:

a) $(3,5 \cdot 10^7) \cdot (4 \cdot 10^8)$ b) $(5 \cdot 10^{-8}) \cdot (2,5 \cdot 10^5)$ c) $(1,2 \cdot 10^7):(5 \cdot 10^{-6})$ d) $(6 \cdot 10^{-7})^2$

24. Calculate, expressing the result as scientific notation, and check your answer by using your calculator:

a) $5,3 \cdot 10^{12} - 3 \cdot 10^{11}$ b) $3 \cdot 10^{-5} + 8,2 \cdot 10^{-6}$ c) $6 \cdot 10^{-9} - 5 \cdot 10^{-8}$ d) $7,2 \cdot 10^8 + 1,5 \cdot 10^{10}$

5. Powers and roots

In this part of the unit, we are going to work with the function "root", which is reciprocal to a power.
For example:

POWERS	ROOTS
A square has a side of 5 cm. What is its area? $A = s^2 = 5^2 = 25$ cm^2.	The area of a square is 49 cm^2. What is the length of its side? s = 7 cm, because $7^2 = 49$. This is a root, a square root: $\sqrt{49} = 7 \iff 7^2 = 49$
A cube has a side of 4 m. What is its volume? $V = s^3 = 4^3 = 64$ m^3.	The volume of a cube is cube 8 cm^3. What is the length of its side? s = 2 cm, because $2^3 = 8$. This is a root, a cubic root: $\sqrt[3]{8} = 2 \iff 2^3 = 8$

Root of a real number **a**, written $\sqrt[n]{a}$ or $a^{1/n}$, where **n** is a natural number, is another real number, **b,** so that
$\sqrt[n]{a} = b \ or \ a^{1/n} \iff b^n = a$

In this expression, **n** is named **index** and **a** is named **radicand**.

25. Calculate the value of the following roots WITHOUT USING A CALCULATOR:

$\sqrt{9} =$ $\sqrt{25} =$ $\sqrt{49} =$ $\sqrt{100} =$ $\sqrt{1} =$

$\sqrt{0} =$ $\sqrt{\dfrac{1}{4}} =$ $\sqrt{\dfrac{1}{9}} =$ $\sqrt{\dfrac{4}{25}} =$ $\sqrt{\dfrac{16}{100}} =$

$\sqrt{0,25} =$ $\sqrt{0,09} =$ $\sqrt{0,0081} =$ $\sqrt{0,49} =$ $\sqrt{7^6} =$

$\sqrt{5^{24}} =$ $\sqrt{2^{10}} =$ $\sqrt{9^{-10}} =$

26. Calculate the value of the following roots WITHOUT USING A CALCULATOR:

$\sqrt[3]{8} =$ $\sqrt[3]{27} =$ $\sqrt[3]{64} =$ $\sqrt[3]{1000} =$ $\sqrt[3]{1331} =$

$\sqrt[3]{-1} =$ $\sqrt[3]{-8}$ $\sqrt[3]{-27} =$ $\sqrt[3]{-1000} =$

$\sqrt[3]{\dfrac{1}{8}} =$ $\sqrt[3]{\dfrac{1}{125}} =$ $\sqrt[3]{\dfrac{64}{125}} =$ $\sqrt[3]{\dfrac{64}{1000}} =$

$\sqrt[3]{0,125} =$ $\sqrt[3]{0,027} =$ $\sqrt[3]{0,001} =$ $\sqrt[3]{-0,216} =$

27. Calculate the value of the following roots BY USING THE DEFINITION OF ROOT:

a) $\sqrt[3]{-8} = -2$ as $(-2)^3 = \boxed{-8}$ **b)** $\sqrt{-8} =$ **c)** $\sqrt[6]{-1} =$ **d)** $\sqrt[5]{-32} =$

e) $\sqrt[4]{81} =$ **f)** $\sqrt{5^2} =$ **g)** $\sqrt[6]{2^6} =$ **h)** $\sqrt{\dfrac{625}{81}} =$

i) $\sqrt[3]{\dfrac{27}{64}} =$ **j)** $\sqrt[4]{-\dfrac{81}{16}} =$ **k)** $\sqrt[5]{3^{15}} =$ **l)** $\sqrt[3]{0,064} =$

m) $\sqrt{0,\overline{1}} =$ **n)** $\sqrt{2,25} =$ **o)** $\sqrt{2,\overline{7}} =$

28. Calculate the value of k:

a) $\sqrt[3]{k} = 2$ **b)** $\sqrt[k]{-243} = -3$ **c)** $\sqrt[5]{k} = \dfrac{2}{3}$ **d)** $\sqrt[k]{1,331} = 1,1$

29. Write as a root:

a) $25^{\frac{1}{3}}$ b) $12^{\frac{1}{4}}$ c) $a^{\frac{3}{5}}$ d) $\left(\dfrac{1}{2}\right)^{\frac{-2}{3}}$ e) $b^{\frac{2}{7}}$ f) $2^{\frac{3}{2}}$

g) $2^{\frac{1}{2}}$ h) $\left(a^2 + b^2\right)^{\frac{1}{2}}$ i) $(a-b)^{\frac{-1}{2}}$ j) $2x^{\frac{3}{5}}$ k) $\left(2x^3 y\right)^{\frac{2}{5}}$ l) $(8x)^{\frac{-2}{3}}$

m) $(-8)^{\frac{-1}{5}}$ n) $-3^{\frac{1}{2}}$

30. Write as a power:

a) $\left(\sqrt[3]{3}\right)^6$ b) $\left(\sqrt[3]{4}\right)^8$ c) $\left(\sqrt{3}\right)^2$ d) $\left(\sqrt[5]{31}\right)^5$ e) $\left(\sqrt{n^2}\right)^7$ f) $\left(\sqrt[3]{m^3}\right)^5$ g) $\left(\sqrt[5]{m^2}\right)^3$ h) $\left(\sqrt[4]{5}\right)^2$

i) $\sqrt[3]{\sqrt{128}}$

31. Use your calculator to find out the value of the following roots, rounding to the thousandths:

a) $\sqrt[4]{8}$ **b)** $\sqrt[5]{9}$ **c)** $\sqrt[6]{25}$ **d)** $\sqrt[3]{10}$

e) $\sqrt[5]{-15}$ **f)** $\sqrt[6]{-40}$ **g)** $\sqrt[4]{2^3}$ **h)** $\sqrt[5]{3^2}$

i) $\sqrt[6]{5^2}$ **j)** $\sqrt[8]{256}$ **k)** $\sqrt[3]{64}$

32. Write the following powers as roots and calculate their values WITHOUT USING YOUR CALCULATOR:

a) $4^{1/2}$ **b)** $125^{1/3}$ **c)** $625^{1/4}$

d) $8^{2/3}$ **e)** $64^{5/6}$ **f)** $81^{3/4}$

g) $8^{-2/3}$ **h)** $27^{-1/3}$

Equivalent roots. Simplification of roots

In roots having a power as radicand, *index* and *exponent* can be *multiplied* or *divided* by a *same number*, in the same way you are used to do it with fractions.
When we divide index and exponent by a same number, we are *simplifying* the root.

As we have seen above, a root can be written as a power, being its exponent a fraction. So, the same operations we use with fractions, can be used with roots (simplifying, ordering, ……..).

Example: Simplify: a) $\sqrt[6]{8}$ b) $\sqrt[12]{625}$ c) $\sqrt[18]{a^{12}}$

Solution (as root):

a) $\sqrt[6]{8} = \sqrt[6]{2^3} = \sqrt[2]{2^1} = \sqrt{2}$, where we have divided by 3.

b) $\sqrt[12]{625} = \sqrt[12]{5^4} = \sqrt[3]{5}$, where we have divided by 4.

c) $\sqrt[18]{a^{12}} = \sqrt[3]{a^2}$, where we have divided by HCF(18, 12) = 6.

Solution (as power):

a) $\sqrt[6]{8} = \sqrt[6]{2^3} = 2^{\frac{3}{6}} = 2^{\frac{1}{2}} = \sqrt{2}$, where numerator and denominator have been divided by 3.

b) $\sqrt[12]{625} = \sqrt[12]{5^4} = 5^{\frac{4}{12}} = 5^{\frac{1}{3}} = \sqrt[3]{5}$, where numerator and denominator have been divided by 4.

c) $\sqrt[18]{a^{12}} = a^{\frac{12}{18}} = a^{\frac{2}{3}} = \sqrt[3]{a^2}$, where numerator and denominator have been divided by HCF(18, 12) = 6.

33. Simplify the following roots:

a) $\sqrt[4]{3^2}$ b) $\sqrt[8]{5^4}$ c) $\sqrt[9]{27}$ d) $\sqrt[5]{1024}$

e) $\sqrt[6]{8}$ f) $\sqrt[9]{64}$ g) $\sqrt[8]{81}$ h) $\sqrt[12]{x^9}$

i) $\sqrt[12]{x^8}$ j) $\sqrt[5]{x^{10}}$ k) $\sqrt[6]{a^2b^4}$ l) $\sqrt[10]{a^4b^6}$

m) $\sqrt[6]{5^3}$ n) $\sqrt[15]{2^{12}}$ o) $\sqrt[10]{a^8}$ p) $\sqrt[12]{x^4y^8z^4}$

q) $\sqrt[8]{\left(x^2y^2\right)^2}$

34. Decide if the following roots are equivalent, and check with your calculator:

a) $\sqrt{5}$, $\sqrt[4]{25}$, $\sqrt[6]{125}$, $\sqrt[8]{625}$ b) $\sqrt{9}$, $\sqrt[3]{27}$, $\sqrt[4]{81}$, $\sqrt[5]{243}$ c) $\sqrt{2}$, $\sqrt[4]{4}$, $\sqrt[6]{8}$, $\sqrt[8]{16}$

35. Rewrite the following roots so that they have the same index and arrange them from lower to higher:

a) $\sqrt{5}$, $\sqrt[5]{2^3}$, $\sqrt[15]{7^2}$ b) $\sqrt[3]{5}$, $\sqrt[5]{7^3}$, $\sqrt[15]{3^2}$ c) $\sqrt[4]{3}$, $\sqrt[6]{16}$, $\sqrt[15]{9}$

d) $\sqrt{2}$, $\sqrt[3]{32}$, $\sqrt[5]{27}$ e) $\sqrt{2}$, $\sqrt[3]{3}$, $\sqrt[4]{4}$, $\sqrt[5]{5}$, $\sqrt[6]{6}$ f) $\sqrt[3]{16}$, $\sqrt[4]{125}$, $\sqrt[6]{243}$

Operations with roots. Laws of roots

Law	Example	Description
$\sqrt[n]{a}\cdot\sqrt[n]{b}=\sqrt[n]{a\cdot b}$	$\sqrt{2}\cdot\sqrt{3}=\sqrt{6}$	**Product** of two roots having the same index is a root having the common index and whose radicand is the product of radicands.
$\sqrt[n]{a}:\sqrt[n]{b}=\sqrt[n]{a:b}$	$\sqrt[5]{12}:\sqrt[5]{4}=\sqrt[5]{3}$	**Quotient** of two roots having the same index is a root having the common index and whose radicand is the quotient of radicands.
$\left(\sqrt[n]{a}\right)^m=\sqrt[n]{a^m}$	$\left(\sqrt[5]{4}\right)^2=\sqrt[5]{16}$	**Power** of a root is another root having the same index and whose radicand is the power of the original radicand.
$\sqrt[m]{\sqrt[n]{a}}=\sqrt[n\cdot m]{a}$	$\sqrt{\sqrt[3]{8}}=\sqrt[6]{8}$	**Root** of a root is another root having the same radicand and whose index is the product of the original index.

Exercises

15

These laws can be easily demonstrated by applying the power rules:

a) $\sqrt[n]{a}\cdot\sqrt[n]{b} = a^{\frac{1}{n}}\cdot b^{\frac{1}{n}} = (a\cdot b)^{\frac{1}{n}} = \sqrt[n]{a\cdot b}$

b) $\sqrt[n]{a} : \sqrt[n]{b} = a^{\frac{1}{n}} : b^{\frac{1}{n}} = (a:b)^{\frac{1}{n}} = \sqrt[n]{a:b}$

c) $\left(\sqrt[n]{a}\right)^{m} = \left(a^{\frac{1}{n}}\right)^{m} = a^{\frac{m}{n}} = \left(a^{m}\right)^{\frac{1}{n}} = \sqrt[n]{a^{m}}$

d) $\sqrt[m]{\sqrt[n]{a}} = \left(a^{\frac{1}{n}}\right)^{\frac{1}{m}} = a^{\frac{1}{n\cdot m}} = \sqrt[n\cdot m]{a}$

Exercises

36. Calculate:

a) $\sqrt{3}\cdot\sqrt{5}\cdot\sqrt{7}$
b) $\sqrt[3]{4}\cdot\sqrt[3]{5}\cdot\sqrt[3]{2}$
c) $\sqrt[5]{12} \div \sqrt[5]{6}$
d) $\sqrt{4} \div \sqrt{2}$

e) $\sqrt[3]{\sqrt{5}}$
f) $\sqrt[6]{\sqrt[3]{6}}$
g) $\sqrt{5}\cdot\sqrt{6}$
h) $\sqrt{2a}\cdot\sqrt{a}$

i) $\sqrt{18} \div \sqrt{50}$
j) $\sqrt{12}\cdot\sqrt{\dfrac{3}{4}}\cdot\sqrt{\dfrac{12}{5}}\cdot\sqrt{\dfrac{15}{4}}$

37. Work out the following products of roots having the same index, and simplify as far as possible:

a) $\sqrt{2}\,\sqrt{32}$
b) $\sqrt{2}\,\sqrt{15}$
c) $\sqrt[3]{3}\,\sqrt[3]{9}$

d) $\sqrt{2}\,\sqrt{8}$
e) $\sqrt{3}\,\sqrt{4}$
f) $\sqrt[3]{2}\,\sqrt[3]{5}$

g) $\sqrt{12}\,\sqrt{6}\,\sqrt{50}$
h) $\sqrt{21}\,\sqrt{7}$
i) $4\sqrt{3}\cdot2\sqrt{27}$

38. Work out the following products of roots having different index, and simplify as far as possible:

a) $\sqrt{2}\,\sqrt[3]{32}$
b) $\sqrt[3]{2}\,\sqrt[4]{8}$
c) $\sqrt[3]{2}\,\sqrt[5]{2}$
d) $\sqrt[3]{9}\,\sqrt[6]{3}$

e) $\sqrt[3]{2^2}\,\sqrt[4]{2}$
f) $\sqrt[4]{a^3}\,\sqrt[6]{a^5}$
g) $\sqrt[3]{2}\,\sqrt{3}\,\sqrt[4]{8}$

39. Simplify, applying, first of all, the laws of roots:

a) $\dfrac{\sqrt{32}}{\sqrt{2}}$
b) $\dfrac{\sqrt{8}}{\sqrt{2}}$
c) $\dfrac{\sqrt[3]{81}}{\sqrt[3]{9}}$
d) $\dfrac{\sqrt{15}}{\sqrt{3}}$
e) $\dfrac{\sqrt{3}}{\sqrt{4}}$

f) $\dfrac{\sqrt[3]{16}}{\sqrt[3]{2}}$
g) $\sqrt{\dfrac{256}{729}}$
h) $\dfrac{\sqrt{21}}{2\sqrt{7}}$
i) $\dfrac{\sqrt{33}}{\sqrt{3}}$
j) $\sqrt[3]{\dfrac{125}{512}}$

k) $\sqrt[4]{\dfrac{16}{625}}$
l) $\dfrac{\sqrt{2}\,\sqrt{8}}{\sqrt{32}}$
m) $\sqrt{\dfrac{154}{9}+23} - \sqrt{4\dfrac{144}{9}}$
n) $\sqrt{\left(-\dfrac{3}{2}\right)^{2}+\left(\dfrac{3\sqrt{3}}{2}\right)^{2}}$

40. Work out the following quotients of roots having different index, transforming them, first of all, into common index. Simplify as far as possible:

a) $\dfrac{\sqrt{8}}{\sqrt[4]{2}}$

b) $\dfrac{\sqrt[3]{9}}{\sqrt[6]{3}}$

c) $\dfrac{\sqrt{2}}{\sqrt[3]{32}}$

d) $\dfrac{\sqrt[4]{4}}{\sqrt[6]{8}}$

e) $\dfrac{\sqrt[3]{7^2}}{\sqrt{7}}$

f) $\dfrac{\sqrt{9}}{\sqrt[3]{3}}$

g) $\dfrac{\sqrt[5]{16}}{\sqrt{2}}$

h) $\dfrac{\sqrt{ab}}{\sqrt[3]{ab}}$

i) $\dfrac{\sqrt[4]{a^3b^5c}}{\sqrt{ab^3c^3}}$

j) $\dfrac{\sqrt[6]{a^3}}{\sqrt[3]{a^2}}$

k) $\dfrac{\sqrt[3]{-2000}}{3\sqrt{2}}$

l) $\dfrac{\sqrt[3]{4}\,\sqrt{3}}{\sqrt[6]{12}}$

m) $\dfrac{\sqrt[8]{8}}{\sqrt[4]{4}\,\sqrt{2}}$

n) $\dfrac{\sqrt[3]{5}\cdot\sqrt{125}}{\sqrt[4]{25}}$

o) $\dfrac{\sqrt[3]{2}\cdot\sqrt{3}\cdot\sqrt[12]{2}}{\sqrt[12]{18}}$

p) $\dfrac{\sqrt[3]{4}\cdot\sqrt{3}\cdot\sqrt[12]{2}}{\sqrt[4]{2}}$

q) $\dfrac{\sqrt[6]{54}\cdot\sqrt[12]{27}}{\sqrt[12]{4}\cdot\sqrt[4]{12}}$

r) $\dfrac{\sqrt[4]{abc^2}\cdot\sqrt[12]{a^3b^5c^2}}{\sqrt[6]{a^2b^2c}}$

41. Simplify:

a) $\left(\sqrt[3]{a^2}\right)^6$

b) $\left(\sqrt[6]{ab^2}\right)^2$

c) $\left(\sqrt{x}\right)^3\cdot\sqrt[3]{x}$

d) $\dfrac{\left(\sqrt[3]{2}\right)^4}{\left(\sqrt[4]{2}\right)^2}$

e) $\dfrac{\sqrt{2}\left(\sqrt[3]{2}\right)^4}{\left(\sqrt[4]{2}\right)^3}$

f) $\sqrt{2}\left(\sqrt[4]{2}\right)^3\left(\sqrt[3]{2}\right)^2$

g) $\dfrac{\left(\sqrt[4]{3}\right)^5}{\left(\sqrt{3}\right)^2\left(\sqrt[3]{3}\right)^4}$

h) $\sqrt{2}\left(\sqrt[4]{2}\sqrt[3]{4}\right)^3$

i) $\sqrt{\sqrt{2^6}}$

j) $\sqrt{\sqrt{12}}$

k) $\left(\sqrt{\sqrt{\sqrt{2}}}\right)^8$

l) $\sqrt[3]{\sqrt[4]{x^5x^7}}$

m) $\sqrt[3]{\sqrt[4]{x^{15}}}$

n) $\left(\sqrt[3]{\sqrt[7]{\sqrt{8x^3}}}\right)^7$

o) $\left(\sqrt{\sqrt[3]{5}}\right)^5\left(\sqrt[4]{5}\right)^3$

p) $\dfrac{\left(\sqrt{x}\right)^3}{\left(\sqrt[3]{\sqrt[4]{x}}\right)^6}$

q) $\dfrac{\left(\sqrt[3]{2}\right)^4\cdot\left(\sqrt[4]{8}\right)^3}{\sqrt{\left(\sqrt[3]{4}\right)^2}}$

r) $\dfrac{\sqrt{\sqrt[3]{a^2}}\cdot\left(\sqrt{a^3}\right)^3}{\left(\sqrt{a}\right)^3\cdot\sqrt[3]{a^4}}$

s) $\dfrac{\left(\sqrt{27}\right)^3\cdot\sqrt{\sqrt[3]{9}}}{\sqrt[3]{81}\cdot\left(\sqrt{3}\right)^3}$

Introducing and extracting factors from roots

These laws allow us to introduce or extract factors from roots.

Example: Introduce factors into the following roots:

Solution: a) $2\sqrt{3}$ b) $2\sqrt[3]{5}$

   a) $2\sqrt{3}=\sqrt{2^2}\cdot\sqrt{3}=\sqrt{4}\cdot\sqrt{3}=\boxed{\sqrt{12}}$

   b) $2\sqrt[3]{5}=\sqrt[3]{2^3}\cdot\sqrt[3]{5}=\sqrt[3]{8}\cdot\sqrt[3]{5}=\boxed{\sqrt[3]{40}}$

Example: Extract from the following roots: a) $\sqrt{200}$ b) $\sqrt[3]{250}$

Solution:

a) $\sqrt{200} = \sqrt{2^3 \cdot 5^2} = \sqrt{2 \cdot 2^2 \cdot 5^2} = 2 \cdot 5 \cdot \sqrt{2} = \boxed{10\sqrt{2}}$

b) $\sqrt[3]{250} = \sqrt[3]{5^3 \cdot 2} = \sqrt[3]{5^3} \cdot \sqrt[3]{2} = \boxed{5\sqrt[3]{2}}$

Exercises

42. Introduce in the roots:

a) $5\sqrt{3}$ b) $2\sqrt[3]{4}$ c) $3\sqrt[4]{2}$ d) $2\sqrt[3]{2}$ e) $2\sqrt{5}$ f) $7\sqrt{a}$

g) $2a\sqrt{3a}$ h) $x\sqrt{\dfrac{1}{x}}$ i) $x^3 y\sqrt{xy}$ j) $\dfrac{1}{3}\sqrt[4]{\dfrac{27}{2}}$ k) $\dfrac{3}{2}\sqrt{\dfrac{2}{3}}$ l) $\dfrac{2}{a}\sqrt{\dfrac{ax}{2}}$

m) $\dfrac{3}{2xy}\sqrt{\dfrac{2xz}{y}}$

43. Extract all possible factors from the following roots:

a) $\sqrt{3^5}$ b) $\sqrt[4]{5^{10}}$ c) $\sqrt[3]{81}$ d) $\sqrt{300}$ e) $\sqrt{18}$ f) $\sqrt{32}$

g) $\sqrt{8}$ h) $\sqrt{75}$ i) $\sqrt{200}$ j) $\sqrt[3]{625}$ k) $\sqrt[4]{32}$ l) $\sqrt[4]{243}$

44. Introduce in the roots:

a) $5\sqrt{3}$ b) $2\sqrt[3]{4}$ c) $3\sqrt[4]{2}$ d) $2\sqrt[3]{2}$ e) $2\sqrt{5}$ f) $7\sqrt{a}$ g) $2a\sqrt{3a}$

h) $x\sqrt{\dfrac{1}{x}}$ i) $x^3 y\sqrt{xy}$ j) $\dfrac{1}{3}\sqrt[4]{\dfrac{27}{2}}$ k) $\dfrac{3}{2}\sqrt{\dfrac{2}{3}}$ l) $\dfrac{2}{a}\sqrt{\dfrac{ax}{2}}$ m) $\dfrac{3}{2xy}\sqrt{\dfrac{2xz}{y}}$

45. Extract all possible factors from the following roots:

a) $\sqrt{\dfrac{27}{4}}$ b) $\sqrt[5]{\dfrac{5x^{10}}{y^8}}$ c) $\sqrt[3]{\dfrac{8x^4 y^3 z}{n^6}}$ d) $3\sqrt{8a^3}$ e) $2x^2 y\sqrt{x^4 y^3}$

f) $\dfrac{xy^2}{3}\sqrt{27xy^3}$ g) $\sqrt{8}$ h) $\sqrt{12}$ i) $\sqrt[3]{16}$ j) $\sqrt[3]{54}$

k) $\sqrt[5]{64}$ l) $\sqrt{12x^3 y^5 z^2}$ m) $\sqrt[3]{\dfrac{8x^4}{81y^6}}$

46. Work out:

a) $\sqrt{\sqrt[3]{b^2}}$ b) $\sqrt{\sqrt{\sqrt{x}}}$ c) $\sqrt[3]{\sqrt{\sqrt[4]{a^3}}}$ d) $\sqrt{x\sqrt{x\sqrt{x}}}$ e) $\sqrt{x\sqrt{x\sqrt{x^2}}}$ f) $\sqrt[3]{4\sqrt{4\sqrt[3]{4}}}$

g) $\sqrt[3]{\sqrt[4]{\sqrt{a^{24}b^{12}c^6}}}$ h) $\sqrt[3]{2\sqrt{(1-a)\sqrt{(1-a)}}}$ i) $\sqrt[4]{\sqrt[3]{\sqrt{\sqrt[3]{2^{144}}}}}$

Like roots

Two or more radicals are said to be **like radicals** if they have the *same index* and the *same radicand*.

For example, these radicals are like radicals $\sqrt[5]{14}$, $8\sqrt[5]{14}$ and $(-2)\sqrt[5]{14}$.

To **add** or **subtract** radicals, they must be like radicals.

Example: Calculate: a) $\sqrt[5]{14} - 8\sqrt[5]{14} + 10\sqrt[3]{14}$

Solution:

a) $\sqrt[5]{14} - 8\sqrt[5]{14} + 10\sqrt[5]{14} = (1 - 8 + 10)\sqrt[5]{14} = \boxed{3\sqrt[5]{14}}$.

b) $6\sqrt[3]{5} - \sqrt[3]{40} = 6\sqrt[3]{5} - \sqrt[3]{2^3 \cdot 5} = 6\sqrt[3]{5} - 2\sqrt[3]{5} = \boxed{4\sqrt[3]{5}}$

c) $\sqrt{2} + \sqrt{8} + \sqrt{18} - \sqrt{32} = \sqrt{2} + \sqrt{2^3} + \sqrt{2 \cdot 3^2} - \sqrt{2^5} = \sqrt{2} + 2\sqrt{2} + 3\sqrt{2} - 4\sqrt{2} = (1 + 2 + 3 - 4)\sqrt{2} = \boxed{2\sqrt{2}}$

Exercises

47. Add the following radicals, extracting previously all possible factors and making them like radicals:

a) $\sqrt{7} + \sqrt{28} - \sqrt{63}$ b) $\sqrt{121} + \sqrt{169} - \sqrt{225}$ c) $\sqrt{2} + \sqrt{8} + \sqrt{18} - \sqrt{32}$

d) $\sqrt{5} + \sqrt{45} + \sqrt{180} - \sqrt{80}$ e) $\sqrt{24} - 5\sqrt{6} + \sqrt{486}$ f) $\sqrt[3]{54} - 2\sqrt[3]{16}$

g) $\sqrt{2} + 3\sqrt{18} - \sqrt{32}$ h) $\sqrt{5} + \sqrt{180} - \sqrt{80}$ i) $\sqrt{2} + \sqrt{32} + 5\sqrt{8}$

j) $3\sqrt{20} + 2\sqrt{5}$ k) $2\sqrt{27} - 4\sqrt{12}$ l) $3\sqrt{28} - 2\sqrt{27}$ m) $\sqrt[3]{16} - \sqrt[3]{54}$

48. Transform into like radicals and simplify:

a) $2a\sqrt{2} - \sqrt{8} + 3\sqrt{2}$ b) $2a\sqrt{3} - \sqrt{27a^2} + a\sqrt{12}$ c) $(3 + a)\sqrt{5} - \sqrt{125} + \sqrt{5a^2}$

d) $4\sqrt{12} - \dfrac{3}{2}\sqrt{48} + \dfrac{2}{3}\sqrt{27} + \dfrac{3}{5}\sqrt{75}$ e) $7\sqrt{54} - 3\sqrt{18} + \sqrt{24} - \dfrac{3}{5}\sqrt{50} - \sqrt{6}$

f) $\sqrt[4]{144} + 3\sqrt{27} - \sqrt{48}$ g) $5\sqrt[6]{256} - 3\sqrt[3]{16} - \sqrt[3]{128}$ h) $\sqrt{3} - \sqrt{108} + \sqrt{648} - \sqrt{1875}$

i) $\dfrac{3\sqrt{3}}{2} - \dfrac{7\sqrt{108}}{4} + \sqrt{648} - \dfrac{2}{3}\sqrt{1875}$

Rationalization

Sometimes, when operating with radicals, there are fractions with radicals at the denominator. **Rationalizing** consists on finding another equivalent fraction, without radicals at the denominator.

To rationalize a fraction, you must multiply **numerator** and **denominator** by a same expression that makes radicals at denominator disappear.

- If denominator is $\sqrt{a}$, we will multiply numerator and denominator by $\sqrt{a}$.

- If denominator is $a + \sqrt{b}$, $a - \sqrt{b}$, $\sqrt{a} + \sqrt{b}$ or $\sqrt{a} - \sqrt{b}$ we will multiply numerator and denominator by their **conjugated** (the same expression, changing the central sign).

Example: Rationalize: a) $\dfrac{5}{\sqrt{3}}$ b) $\dfrac{3}{2\sqrt{2}}$ c) $\dfrac{3}{\sqrt{2}-1}$

Solution:

a) $\dfrac{5}{\sqrt{3}}$ $\rightarrow$ We must multiply numerator and denominator by $\sqrt{3}$ $\rightarrow$ $\dfrac{5\cdot\sqrt{3}}{\sqrt{3}\cdot\sqrt{3}}=\dfrac{5\cdot\sqrt{3}}{(\sqrt{3})^2}=\boxed{\dfrac{5\cdot\sqrt{3}}{3}}$

b) $\dfrac{3}{2\sqrt{2}}$ $\rightarrow$ We must multiply numerator and denominator by $\sqrt{2}$ $\rightarrow$ $\dfrac{3\cdot\sqrt{2}}{2\cdot\sqrt{2}\sqrt{2}}=\dfrac{3\cdot\sqrt{2}}{2\cdot(\sqrt{2})^2}=\boxed{\dfrac{3\cdot\sqrt{2}}{4}}$

c) $\dfrac{3}{\sqrt{2}-1}$ $\rightarrow$ We must multiply numerator and denominator by $(\sqrt{2}+1)$ $\rightarrow$

$\rightarrow$ $\dfrac{3\cdot(\sqrt{2}+1)}{(\sqrt{2}-1)\cdot(\sqrt{2}+1)}=\dfrac{3\cdot(\sqrt{2}+1)}{(\sqrt{2})^2-1}=\dfrac{3\cdot(\sqrt{2}+1)}{2-1}=\boxed{3\sqrt{2}+3}$

Exercises

49. Rationalize the following expressions:

a) $\dfrac{2}{\sqrt{3}}$ b) $\dfrac{1}{3\sqrt{2}}$ c) $\dfrac{5}{\sqrt{5}}$ d) $\dfrac{2\sqrt{6}}{\sqrt{2}}$ e) $\dfrac{\sqrt{27}}{\sqrt{8}}$ f) $\sqrt{\dfrac{2}{3}}$ g) $\sqrt{\dfrac{5}{2}}$ h) $\dfrac{3}{\sqrt{5}}$

50. Multiply by its conjugated each of the following expressions:

a) $2-\sqrt{x}$ b) $\sqrt{x}-2$ c) $\sqrt{x}+\sqrt{y}$

51. Rationalize the following expressions:

a) $\dfrac{2}{3-\sqrt{5}}$ b) $\dfrac{8\sqrt{5}}{\sqrt{7}-\sqrt{3}}$ c) $\dfrac{9}{\sqrt{x}-\sqrt{y}}$ d) $\dfrac{1+\sqrt{x}}{1-\sqrt{x}}$

53. Rationalize the following expressions:

a) $\dfrac{2}{1-\sqrt{3}}$ b) $\dfrac{5}{\sqrt{7}-\sqrt{2}}$ c) $\dfrac{\sqrt{2}}{5-\sqrt{2}}$ d) $\dfrac{\sqrt{3}}{\sqrt{12}+\sqrt{2}}$

54. Rationalize and simplify:

a) $\dfrac{2}{1+\sqrt{2}}$ b) $\dfrac{4}{3-\sqrt{2}}$ c) $\dfrac{23}{5-\sqrt{2}}$ d) $\dfrac{1}{1-\sqrt{3}}$ e) $\dfrac{1}{\sqrt{5}+3}$

f) $\dfrac{1}{\sqrt{3}-\sqrt{2}}$ g) $\dfrac{10}{\sqrt{3}+\sqrt{2}}$ h) $\dfrac{\sqrt{2}}{\sqrt{2}+3}$ i) $\dfrac{1+\sqrt{3}}{1-\sqrt{3}}$

---

6. Percentages

A **percentage** is a ratio of a number to 100. A percentage is expressed using the symbol %.

A percentage is also equivalent to a fraction with a denominator of 100.

For example, 65% is equivalent to the fraction $\dfrac{65}{100}$.

To solve problems involving percentages, we will use the following expression:

$$Portion = \frac{Percentage}{100} \cdot Whole$$

Anyway, problems with percentages can also be solved as a direct proportionality. We are seeing it in the following examples.

6.1. Calculating the portion

Example: What is the 35% of 300?

Solution: We will solve it by applying the percent expression and as a direct proportionality.

Method 1: Percentage expression:	**Method 2:** As a direct proportionality:
- Portion = x? Calculate: - Percentage = 35 - Whole = 300 $\quad x = \frac{35 \cdot 300}{100} = \boxed{105}$. $x = \frac{35}{100} \cdot 300$	**Whole:** $\quad$ $\boxed{D}$ $\quad$ **Portion:** $\quad$ 100 $\qquad\qquad$ 35 $\quad$ 300 $\qquad\qquad$ x $\frac{100}{300} = \frac{35}{x} \rightarrow x \cdot 100 = 300 \cdot 35; \; x = \frac{35 \cdot 300}{100} = \boxed{105}$.

<table>
<tr><td rowspan="2">Exercises</td><td>55. Calculate:</td></tr>
<tr><td>a) 60% of 80   b) 30% of 91   c) 55% of 72   d) 130% of 90</td></tr>
</table>

6.2. Calculating the whole

Example: 54% of what number is 108?

Solution: We will solve it by applying the percent expression and as a direct proportionality.

Method 1: Percentage expression:	**Method 2:** As a direct proportionality:
- Portion = 108 $\quad$ Isolate x: - Percentage = 54 - Whole = x? $\quad 108 \cdot \frac{100}{54} = x = \frac{10800}{54} =$ $108 = \frac{54}{100} \cdot x \qquad = \boxed{200}$.	**Whole:** $\quad$ $\boxed{D}$ $\quad$ **Portion:** $\quad$ 100 $\qquad\qquad$ 54 $\quad$ x $\qquad\qquad$ 108 $\frac{100}{x} = \frac{504}{108} = \frac{1}{2} \rightarrow x \cdot 1 = 2 \cdot 100 = \boxed{200}$.

56. Calculate and complete:

a) 80% of _____ is 32.

b) 39% of _____ is 39.

c) 497.8 is 76% of ___ .

d) 162 is 18% of ___.

e) 39% of ___ is 105.3.

f) 73% of ___ is 83.22.

g) 124% of ___ is 328.6.

h) 205% of ___ is 750.3.

All together. Calculating percentages, portions and wholes

57. What percentage of 85 is 41? Round to the nearest tenth of a percent.

58. What percentage of 666 is 247? Round to the nearest tenth of a percent

59. Eighty-two is 24.8% of what number? Round to the nearest hundredth.

60. Forty-one is 35.2% of what number? Round to the nearest hundredth.

61. Thirty-two and seven tenths percent of 695 is what number?

62. Seventy-three and twelve hundredths percent of 35 is what number?

63. Thirty-seven is what percentage of 156? Round to the nearest tenth of a percent.

64. Two hundred thirty-two is what percentage of 124? Round to the nearest tenth of a percent.

65. The cost of a certain model of Ford is 120% of what it was 5 years ago. If the cost of the automobile 5 years ago was $20,400, what is the cost today?

66. In a 1st ESO group, 10 students played football, 15 basketball and 5 tennis. What is the percentage of the students playing football?

67. In a study of 615 people, 185 said they jog for exercise. What percentage of those surveyed jog? Round to the nearest whole percent.

68. Based on a survey, approximately 77% of TV sets in the United States receive cable. If there are about 304,000,000 sets in the United States, how many do not get cable? Round to the nearest million.

69. The town of Verboort has a population of 17,850, of which 48% is male. Of the men, 32% are 40 years or older. How many men are there in Verboort who are younger than 40?

6.3. Percentage changes

As you have seen, percent changes can be calculated by using previous expression. Anyway, there is an easier expression you can use. It is the following:

Expression to calculate percentage changes: $F.A. = I.A. \cdot \left(1 \pm \dfrac{\%}{100}\right)$

Where: - F.A.: Final amount.

- I.A.: Initial amount.

- %: Percentage change.

- $\pm$: Use (+) when you consider an increase and a (-) in decreases.

Example: Prices of houses have increased 20% from five years ago. In that moment, I paid 120,000 euros for my house. How much would I have to pay today?

Solution:

% = 20; I.A. = 120,000; F.A. = x?; plus or minus? As prices have increased, we use <u>plus</u> →

$$F.A. = 120000 \cdot \left(1 + \dfrac{20}{100}\right) = 120000 \cdot (1 + 0.2) = 120000 \cdot 1.2 = \boxed{144,000 \text{ euros}}.$$

Example: In the shop where I am used to buying my clothes, there are 15% sales today. I have paid 25.5 euros for T-shirt. What was its price before sales?

Solution:

% = 15; F.A. = 25.5; I.A. = x?; plus or minus? As prices have decreased, we use <u>minus</u> →

$$F.A. = I.A. \cdot \left(1 \pm \dfrac{\%}{100}\right) \rightarrow 25.5 = I.A. \cdot \left(1 - \dfrac{15}{100}\right); \quad 25.5 = I.A. \cdot 0.85; \quad \dfrac{25.5}{0.85} = I.A. = \boxed{30 \ euros}$$

Example: When John began to work in his company, his salary was 1100 euros per month. From that moment, his salary has increased twice, first, 15%, and later, 20%. But, this year, as there are problems with Spanish economy, he has had his salary decreased 35%. What is his salary today?

Solution: We might think *"it increased 15 + 20 = 35%, and it decreased 35%, so, he should have the same initial salary"*. We are going to see this is not true. As there are more than one percent change, we will use more than one parentheses in above expression:

$$F.A. = 1100 \cdot \left(1 + \dfrac{20}{100}\right) \cdot \left(1 + \dfrac{15}{100}\right) \cdot \left(1 - \dfrac{35}{100}\right) = 1100 \cdot 1.2 \cdot 1.15 \cdot 0.65 = \boxed{986.7 \text{ euros}}.$$

You can see his salary has decreased. Be careful in the future, don't forget this!!!

70. I am taking part in a program to lose weight. I have already lost a 20%. If my weight now is 60 kg, what was my weight before the program?

71. Today there are 20% sales in a shop near my house. This was a T-shirt's label: "Before 25 €; Today 22.5 €". Do you think this label is correct? Why?

72. During ten last years, prices of houses, first increased 10%, later increased 20% and, this year, they have decreased 30%. Ten years ago, I paid 150,000 € for my house. Check if its price has changed from them. Has it increased or decreased?

7. Logarithms

Logarithms are related to powers. For example, if you wonder "*¿What is the exponent of a power of 5 to obtain 25?*" That is the idea of logarithm: $\log_5 25 = 2$ which is read as: "*the base 5 logarithm of 25 is 2*".

Observe this is because $5^2 = 25$.

Definition of logarithm

$$\log_a b = n \qquad \Leftrightarrow \qquad a^n = b$$

An easy way to calculate logarithms is previously expressing the number as a power having as base the base of the logarithm. At that moment, the logarithm will be the exponent of that power.

Examples:

a) $\log_3 81 = \log_3 3^4 = 4$ b) $\log_{10} 0,001 = \log_{10} 10^{-3} = -3$ c) $\log_2 0,25 = \log_2 \dfrac{1}{4} = \log_2 2^{-2} = -2$

73. By using the definition, calculate the following logarithms:

a) $\log_3 9$	e) $\log_2 \sqrt{2}$	i) $\log_4 64$	m) $\log_4 256$	q) $\log_2 1024$
b) $\log_3 81$	f) $\log_2 \sqrt{8}$	j) $\log_{10} 0,01$	n) $\log_4 1/64$	r) $\log_2 1/64$
c) $\log_3 1/9$	g) $\log_{10} 1000$	k) $\log_4 1/16$	o) $\log_2 0,125$	s) $\log_3 \sqrt{27}$
d) $\log_3(-9)$	h) $\log_4 2$	l) $\log_5 0,2$	p) $\log_4 1$	t) $\log_2 \log_2 4$

Sol.: a) 2; b) 4; c) -2; d) $\nexists$; e) 1/2; f) 3/2; g) 3; h) 1/2; i) 3; j) -2; k) -2; l) -1; m) 4; n) -3;

o) -3; p) 0; q) 10; r) -6; s) 3/2; t) 1)

Decimal logarithms

Base 10 logarithms are named as *deciml logarithms*, and they are not expressed as $\log_{10}$, but simply as *log*

.

$$\log_a 1 = 0 \quad \Leftrightarrow \quad a^0 = 1$$
$$\log_a a = 1 \quad \Leftrightarrow \quad a^1 = a$$
$$\log_a 0 = No\, existe \quad \Leftrightarrow \quad a^b \neq 0$$

Exercises

74. Calculate (without calculator) the base 10 logrithms of the following numbers and check your results with your calculator:

   a) 10.000   b) 1.000.000   c) 0,001   d) 1/1.000.000   e) 108

   f) 10^{-7} g) 10   h) 1

Natural or neperian logarithms

Base *e* logarithms (*e = 2,718...*), are named as natural, naperian or napierian logarithms, due to John Napier, an English mthematician. They are expressed as *ln*.

Exercises

75. By using the definition, calculate the value of x in the following expressions:

   a) $\log_2 8 = x$ b) $\log_2 1/8 = x$ c) $\log 100 = x$ d) $\log_3 x = 3$ e) $\ln x = 2$

   f) $\log_3 x = -2$ g) $\log x 49 = 2$ h) $\log_x 8 = 3$ i) $\ln e^3 = x$ j) $\log_x 64 = 1$

   k) $\log_x 25 = -1$ l) $\lg_{1/100} 100 = x$ m) $\log_x 0.01 = 2$ n) $\ln x = -1/2$ o) $\log_{1/36} x = 2$

   p) $\log_x 2 = 0$ q) $\log_{0.25} x = 2$ r) $\log_2 (-16) = x$ s) $\log_x 125 = -3$

   Sol.: *a) 3; b) -3; c) 2; d) 27; e) e^2; f) 1/9; g) 7; h) 2; i) 3; j) 64; k) 1/25; l) -1; m) 0,1; n) $\sqrt{e}/e$; o) 1/1296; p) DNE; q) 0,0625; r) DNE; s) 1/5*

Laws of logarithms

$$\log(A \cdot B) = \log A + \log B$$
$$\log(A/B) = \log A - \log B$$
$$\log(A^n) = n \cdot \log A$$

76. By using the previous expressions, calculate:

a) $\log_6 \dfrac{1}{36}$

b) $\log_3 \sqrt[4]{27}$

c) $\log_3 \dfrac{\sqrt{243}}{3}$

d) $\log_a \dfrac{1}{\sqrt{a}}$

e) $\ln e^2$

f) $\log_4 \dfrac{1}{\sqrt[5]{64}}$

g) $\log_3 \sqrt[3]{9}$

h) $\ln \dfrac{1}{e}$

i) $\log_4 2$

j) $\log_8 2$

k) $\log_8 \sqrt{32}$

l) $\ln \sqrt[3]{e}$

m) $\log_2 64$

n) $\log_4 \dfrac{1}{64}$

o) $\log_3 \dfrac{3}{\sqrt[5]{81}}$

p) $\log_3 \dfrac{\sqrt{3}}{9}$

q) $\ln \dfrac{\sqrt{e}}{e}$

r) $\log_4(-4)$

s) $\log_2 \sqrt[3]{32}$

t) $\log_3 \sqrt{27}$

u) $\log_2 \dfrac{\sqrt[5]{64}}{8}$

v) $\ln \dfrac{1}{\sqrt[3]{e^2}}$

Sol.: *a) -2; b) 3/4; c) 3/2; d) -1/2; e) 2; f) -3/5; g) 2/3; h) -1; i) 1/2; j) 1/3; k) 5/6; l) 1/3; m) 6; n) -3; o) 1/5; p) -3/2; q) -1/2; r) No existe; s) 5/3; t) 3/2; u) -9/5; v) -2/3.*

77. Express as a function of *log2* and check with your calculator:

a) 16   b) 5   c) 32/5   d) 0,25   e) 0,625   f) 250   g) 1/40

h) $\sqrt[3]{16}$ i) 16/5   j) 0,32   k) 0,08   l) $\sqrt[5]{80}$ m) $\sqrt[3]{0,08}$

Sol.: *a) 4log2; b) 1-log2; c) -1+6log2; d) -2log2; e) 1-4log2; f) 3-2log2; g) -1-2log2; h) $\dfrac{4}{3}\log 2$; i) -1+5log2; j) -2+5log2; k) -2+3log2; l) $\dfrac{1}{5}(1+3\log 2)$ m) $-\dfrac{2}{3}+\log 2$*

78. Express as a function of *ln2*:

a) $\ln 8$ b) $\ln \dfrac{e}{2}$ c) $\ln \dfrac{e^3}{4}$ d) $\ln \dfrac{4}{\sqrt{e}}$ e) $\ln \sqrt{2e}$

Sol.: *a) 3ln2; b) 1-ln2; c) 3-2ln2; d) $-\dfrac{1}{2}+2\ln 2$; e) $\dfrac{1+\ln 2}{2}$*

Review exercises

1. Classify the following numbers into rational or irrational numbers:

$\dfrac{1}{8}$ $\dfrac{\pi}{3}$ $\sqrt{5}$ $2,6$ 0 -3 $-\dfrac{25}{3}$ $\sqrt{13}$ $0,1$ $6,\widehat{4}$ 534 $1,414213...$

2. Classify the following numbers into rational or irrational numbers:

$\dfrac{\pi}{2}$ $\sqrt{3}$ $\sqrt{4}$ $0,0015$ -10 $\dfrac{5}{6}$ $2,\overline{3}$ $2,020020002...$

3. Point out which of the following numbers are irrational and why:

a) $3,629629629....$ **d)** $0,123456789...$ **g)** $0,130129128...$

b) $0,128129130...$ **e)** $7,129292929...$

c) $5,216968888...$ **f)** $4,101001000...$

4. Work out, directly and writing first as fractions, and check you obtain the same results:

k) $0,\widehat{6}:0,0\widehat{5}+0,25$ **l)** $1,25-1,1\widehat{6}+1,\widehat{1}$ **m)** $2,\overline{7}\cdot1,8+2,2\overline{6}:0,11\overline{3}$

n) $1,9\widehat{2}+0,25(0,2\widehat{5}+0,\widehat{5})$ **o)** $\sqrt{2,\overline{7}}$ **p)** $0,8\widehat{3}-0,8:0,\widehat{6}$

q) $4,08\widehat{3}\cdot11,\widehat{1}-0,1\widehat{5}:0,3$ **r)** $0,\widehat{6}+1,3\widehat{8}\cdot0,72$ **s)** $0,\widehat{5}-0,1\widehat{5}+1,2\widehat{3}$

5. Fill in the next chart:

GRAPHICAL REPRESENTATION	INTERVAL	Math. expression
–1 3	$[-1,3]$	$\{x\in IR/\ -1\leq x\leq3\}$
0 2		
–2 4		
	$[-2,1)$	
		$\{x\in IR/\ 1<x\leq5\}$
–1 ∞		
		$\{x\in IR/\ x<2\}$
	$(0,\infty)$	

6. Fill in the next chart:

GRAPHICAL REPRESENTATION	INTERVAL	Math. expression
	(-1,5)	
		{x∈ R/ x≤0}
	[2/3,∞)	
		{x∈ IR/ -2<x≤2}
		{x∈ IR/ |x|<3}
		{x∈ IR/ |x|≥3}
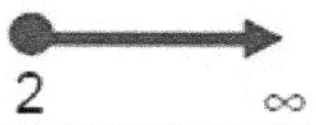 ●——→ 2 ∞		
	[-1,1]	
		{x∈ IR/ x<-1}

7. Fill in the next chart:

GRAPHICAL REPRESENTATION	INTERVAL	Math. expression
●————————● -4 4		
	(-∞,-2)U(2,∞)	
	(-∞,2)U(2,∞)	
		{x∈ IR/ |x|≤5}
	[-2,2]	
 ○——○ -3 3		

28

8. Work out the union and the intersection of the following sets. First of all, draw them:

a) A=[-2,5)
    B=(1,7)
b) C=(-1,3]
    D=(1,6]

c) E=(0,3]
    F=(2,∞)
d) G=(-∞,0]
    H=(-3,∞)

e) I=[-5,-1)
    J=(2,7/2]
f) K=(-∞,0)
    L=[0,∞)

g) M=(2,5)
    N=(5,9]
h) O=[-3,-1)
    P=(2,7]

i) Q=(-3,7)
    R=(2,4]

j) S=[-3,2)
    T=(0,∞)
    U=[1,4]

9. Work out:

a) $\dfrac{7}{4} - \left(\dfrac{5}{3} + \dfrac{2}{3} \cdot \dfrac{1}{5}\right) + 2$

b) $1 - \dfrac{3}{5}\left(\dfrac{2}{3} + \dfrac{1}{2}\right)$

c) $3 - \dfrac{4}{5} : 2 + \dfrac{1}{2} \cdot \left(1 - \dfrac{14}{3}\right)$

d) $\dfrac{5}{6} : \left(\dfrac{2}{3} + 1\right) - \dfrac{3}{4}\left(\dfrac{2}{3} - 1\right)$

e) $\dfrac{\dfrac{7}{5} - \dfrac{3}{4} \cdot \dfrac{2}{5}}{3 - \dfrac{1}{4}}$

f) $\dfrac{\left(\dfrac{1}{4} - \dfrac{7}{8}\right) : \dfrac{2}{3} + 1}{\dfrac{5}{6} \cdot \left(\dfrac{2}{3} - \dfrac{3}{4}\right)}$

g) $\dfrac{\left(-\dfrac{3}{4}\right) \cdot \left(\dfrac{8}{9}\right)}{\dfrac{5}{3} : \dfrac{7}{6}}$

10. Work out:

a) $-\left(\dfrac{3}{4} - 1\right) - \left(\dfrac{1}{2} - \dfrac{1}{4} + \dfrac{1}{5} - 2\right)$

b) $3 - \dfrac{1}{4} + \left(-2 - \dfrac{1}{2} + \dfrac{3}{5}\right)$

c) $-\left(\dfrac{1}{2} - \dfrac{1}{4}\right) - \left(3 - \dfrac{1}{2} + \dfrac{5}{3} - 1\right)$

d) $\dfrac{1}{2} + 3 - \left(-\dfrac{1}{4} + \dfrac{5}{2} - 8 + 1\right)$

e) $-\left(\dfrac{4}{5} - 1\right) - \left(2 - \dfrac{3}{2} + 3 - \dfrac{4}{5}\right)$

f) $-\left(-1 + \dfrac{3}{7} - 3 + \dfrac{4}{3}\right) - \left(-3 + \dfrac{1}{2}\right)$

g) $\dfrac{1 - \dfrac{3}{4} + \dfrac{1}{2}}{\dfrac{2}{3} + 3}$

h) $\dfrac{\left(1 - \dfrac{4}{5}\right) - (-2 + 1)}{\dfrac{4}{5} - 3 + \dfrac{1}{2}}$

i) $\dfrac{29}{7} - \left(2 - \dfrac{4}{5}\right) : \left(\dfrac{3}{5} + \dfrac{1}{2} - \dfrac{3}{4}\right)$

j) $\dfrac{5}{6} - \dfrac{3}{7} : \dfrac{9}{14} + \left(\dfrac{2}{3} - \dfrac{4}{9}\right) : \dfrac{16}{45} - \dfrac{1}{24}$

11. Calculate, expressing the result as scientific notation, and check your answer by using your calculator:

a) $(2,8 \cdot 10^{-5}):(6,2 \cdot 10^{-12})$ 　　 b) $(7,2 \cdot 10^{-6})^3:(5,3 \cdot 10^{-9})$ 　　 c) $7,86 \cdot 10^5 - 1,4 \cdot 10^6 + 5,2 \cdot 10^4$

d) $(3 \cdot 10^{-10} + 7 \cdot 10^{-9}):(7 \cdot 10^6 - 5 \cdot 10^5)$

12. Calculate, expressing the result as scientific notation, and check your answer by using your calculator:

a) $2,5 \cdot 10^7 + 3,6 \cdot 10^7$ 　　 b) $4,6 \cdot 10^{-8} + 5,4 \cdot 10^{-8}$ 　　 c) $1,5 \cdot 10^6 + 2,4 \cdot 10^5$ 　 d) $2,3 \cdot 10^9 + 3,25 \cdot 10^{12}$

e) $3,2 \cdot 10^8 - 1,1 \cdot 10^8$ 　　 f) $4,25 \cdot 10^7 - 2,14 \cdot 10^5$ 　　 g) $7,28 \cdot 10^{-3} - 5,12 \cdot 10^{-3}$ 　 h) $(2 \cdot 10^9) \cdot (3,5 \cdot 10^7)$

i) $(2 \cdot 10^5)^2$ 　　 j) $(1,4 \cdot 10^{15} + 2,13 \cdot 10^{18}) \cdot 2 \cdot 10^{-5}$ 　　 k) $2,23 \cdot 10^{-3} + 3 \cdot 10^{-4} - 5 \cdot 10^{-5}$

l) $(0,55 \cdot 10^{23} - 5 \cdot 10^{21}) \cdot 2 \cdot 10^{-13}$

29

13. Calculate the volume (m^3) of the Earth, if its average radius is 6378 km, giving the result with only two digits.

14. One type of human cell having a cylinder shape, has a diameter of about 7 millionths of a meter and about 2 millionths of a meter in height. Calculate its volume in scientific notation.

15. With a laboratory instrument, the weight of one hundred rice seeds has been determined and it is 0,0000277 kg. How many of these seeds are there in 1000 ton of rice?

16. The light from the Sun takes 8 minutes and 20 seconds to come to the Earth. Calculate the distance Sun-Earth.

17. Simplify:

a) $\dfrac{\sqrt{12}}{\sqrt{3}}$ b) $\dfrac{\sqrt[3]{4}}{\sqrt{2}}$ c) $\sqrt[4]{\dfrac{5}{12}}:\sqrt[4]{\dfrac{20}{3}}$ d) $\dfrac{\sqrt[4]{a^2}}{\sqrt[4]{a}}$ e) $\sqrt{\dfrac{3}{2}}:\sqrt{\dfrac{2}{3}}$ f) $\dfrac{\sqrt[6]{20}}{\sqrt[4]{10}}$

g) $\sqrt[3]{2^2}\cdot\sqrt[4]{2}$ h) $\sqrt[4]{a^3}\cdot\sqrt[6]{a^5}$ i) $\dfrac{\sqrt[8]{8}}{\sqrt[4]{2}\cdot\sqrt{2}}$

18. Extract all possible factors from the following roots:

a) $\sqrt{900}$ b) $\sqrt[3]{64x^9y^4}$ c) $\sqrt[4]{8^2x^{10}y^6z^8}$

19. Introduce factors in these roots:

a) $3\sqrt[3]{5}$ b) $2^2\sqrt[4]{3}$ c) $\dfrac{3}{5}\sqrt[5]{\dfrac{5}{3}}$

20. Calculate:

a) $\sqrt[4]{x^2y^2}\cdot\sqrt{xz}\cdot\sqrt{yz}$ b) $\left(2+\sqrt{5}\right)\left(3-\sqrt{5}\right)$ c) $2a\sqrt{a}\cdot ab^2\sqrt[3]{b^2}$ d) $\dfrac{1}{3}\sqrt{\dfrac{a}{b}}\cdot 6\sqrt[3]{ab^2}\cdot\sqrt[4]{\dfrac{a^2}{b}}$

e) $4\sqrt{72}:\sqrt{8}$ f) $\sqrt[4]{xy^2z^3}:\sqrt[6]{x^2z^3}$ g) $\dfrac{\sqrt[3]{ab^2c^2}}{\sqrt[4]{a^2bc}}$ h) $\dfrac{\sqrt{2a}}{\sqrt[3]{a}}$

i) $28\sqrt{x^4y^3}:7\sqrt{x^3y}$ j) $\dfrac{\sqrt[4]{x^3y^2z^3w}}{\sqrt[3]{x^2y^2zv^2}}$

21 Calculate:

a) $\sqrt{63}-\dfrac{5}{2}\sqrt{28}+\sqrt{112}$ b) $\sqrt{3}+3\sqrt{3}-5\sqrt{3}$ c) $2\sqrt{8}+4\sqrt{72}-7\sqrt{18}$

d) $3\sqrt{2}+4\sqrt{8}-\sqrt{32}+\sqrt{50}$ e) $5\sqrt{12}+\sqrt{27}-8\sqrt{75}+\sqrt{48}$ f) $\sqrt{2}+\dfrac{3\sqrt{2}}{4}-\dfrac{5\sqrt{2}}{3}$

22. Calculate:

a) $\sqrt{320}+\sqrt{80}-\sqrt{500}$ b) $\sqrt{125}+\sqrt{54}-\sqrt{45}$ c) $\sqrt[3]{40}+\sqrt[3]{135}-\sqrt[3]{5}$

23. Calculate:

a) $\dfrac{2\sqrt{3}-5}{\sqrt{3}-2}$

b) $\dfrac{1+\sqrt{3}}{1-\sqrt{3}}$

c) $\dfrac{\sqrt{5}+2\sqrt{3}}{2\sqrt{5}-\sqrt{3}}$

d) $\dfrac{3\sqrt{2}-4}{3\sqrt{2}+4}$

e) $\dfrac{2\sqrt{8}-3\sqrt{2}}{2\sqrt{8}+3\sqrt{2}}$

f) $\dfrac{\sqrt{5}+\sqrt{3}}{\sqrt{5}-\sqrt{3}}$

g) $\dfrac{3\sqrt{5}-4}{\sqrt{5}-2}$

h) $\dfrac{24-13\sqrt{3}}{2\sqrt{3}-3}$

i) $\dfrac{2\sqrt{2}}{\sqrt{3}-\sqrt{2}}$

j) $\dfrac{4-\sqrt{6}}{\sqrt{6}-2}$

k) $\dfrac{2-\sqrt{8}}{2+\sqrt{2}}$

l) $\dfrac{-\sqrt{3}-1}{1-\sqrt{3}}$

m) $\dfrac{9+4\sqrt{3}}{3\left(4-\sqrt{3}\right)}$

n) $\dfrac{\sqrt{2}+4}{2-\sqrt{2}}$

o) $\sqrt{x}+\dfrac{x}{2\sqrt{x}}$

24. WITHOUT OPERATING, Insert the decimal point in the result of fthe following operations:

a) $23.64 + 17.125 = 40765$

b) $15.92 - 7.874 = 8046$

c) $13.9 \cdot 0.5 \cdot 2.7 = 171665$

d) $0.42 + 8.5 = 892$

e) $3.28 \cdot 4.15 = 13612$

f) $123.25 : 2.9 = 425$

g) $4.025 - 3.15 = 875$

h) $16.5 \cdot 2.08 = 3432$

i) $3.9285 : 0.45 = 873$

25. The population of a town is 652 000 and 35% of it live in the central district. How many of the people live in this district?

26. During a sale the price of a television set is 150 € which is 75 % of the usual price, what is the original price?

27. 6% of the population in Murcia are immigrants and there are 9900 immigrants living in the city, what is the population in Murcia?

28. Some clothes are priced at 68 € and there is a discount of 7%, what is the final price?

29. I have bought a pair of jeans for 33 €, the VAT is 10 %, what was the price before?

30. The price of an electric oven before taxes is 560 € plus 17% VAT and the salesman offers a 12% discount, what is the final price?

31. Last year there were 1560 employees in a company, this year 312 new people have been employed. What has been the percentage of increase in the staff of the company?

32. I have paid 161 € for a coat and the original price was 230 €. What is the % discount?

33. In shop A, a mobile phone costs 99 € plus VAT (16%). In shop B, the same model costs 114.90 € (VAT included). In what shop is the mobile phone cheaper?

34. At a sporting event that brings together 750 athletes, 30% of them are American, 8% Asian, 16% African, and the remainder European. How many European athletes take part in the match?

35. What's the total number of guests who attend a wedding, knowing that there are 33 men and 45% are women?

36. The price of a book after an increase of 20 % is 4.20 €. How much did it cost before the rise?

37. Express as a function of log 2 and log 3 the following logarithms and chek with your calculator:

   a) log 25   b) log 24   c) log 4/3   d) log 9/4   e) $\log \sqrt[3]{6}$ f) log 30   g) log 162

   h) log 3,6   i) log 1,2   j) log 90   k) log 0,27   l) log 0,72   m) $\log \sqrt{3,6}$

> ***Sol:*** *a) 2-2log2; b) 3log2+log3; c) 2log2-log3; d) 2log3-2log2; e) $\frac{\log 2 + \log 3}{3}$; f) 1+log3;*
>
> *g) log2+4log3; h) -1+2log2+2log3; i) -1+2log2+log3; j) 1+2log3; k) -2+3log3;*
>
> *l) -2+3log2+2log3; m) -1/2+log2+log3*

38. Knowing that *log7,354* = 0,866524..., calculate (without calculator):

   a) log 735,4   b) log 0,007354   c) log 7354

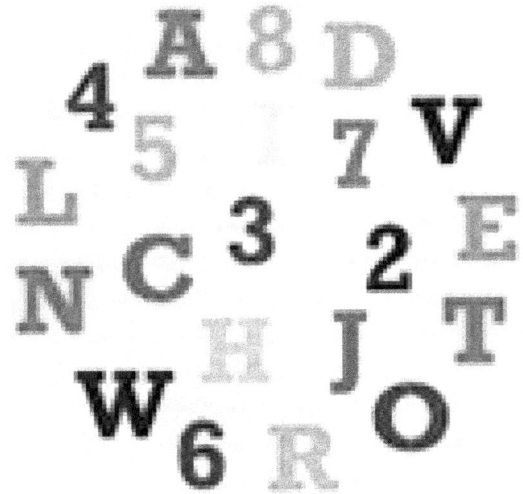

Unit 2.- Polynomials

1. Algebraic expressions. Monomials

An algebraic expression is a combination of numbers, variables (algebraic quantities) and arithmetic operators: $+, -, x, \div$.　　　For example: $2ax + 5b - 3x2y$

Terms

In the following algebraic expressions, the terms are separated by "+" or "–" signs.

Examples:

$4x^2 - 2x + 1$　　has three terms　　　　　　　　　$6x + 7$　　　　has two terms

Monomials and operations

A monomial consists of the product of a known number **(coefficient)** by one or several letters with exponents that must be constant and positive whole numbers **(literal part)**.

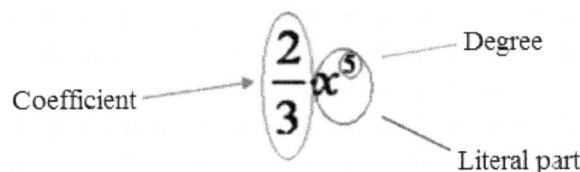

Two or more monomials are said to be **like monomials** if they have the same literal part.

Example: The expressions $21x^2$, $8x^3$, $\frac{2}{5}x^2$, $-3x^5$ are monomials, and $21x^2$ and $\frac{2}{5}x^2$ are *like monomials*.

Addition and subtraction of monomials

Two monomials can only be added or subtracted if they are **like monomials**. So, they must have the same literal part. In this case, we add or subtract the coefficients and we **leave the literal part unchanged**. When the literal part is different, the addition is left indicated.

The result of the **addition** or **subtraction** two **like monomials** is another like monomial whose coefficient is the sum or subtraction of the coefficients.

$$ax^n \pm bx^n = (a \pm b)x^n.$$

1. Calculate:

a) $3x^2 + 4x^2 - 5x^2$

b) $7xy^2z - 2xy^2z + xy^2z - 6xy^2z$

c) $x^4 + x^2 - 3x^2 + 2x^4 - 5x^4 + 8x^2$

d) $3a^2b - 5ab^2 + a^2b + ab^2$

e) $\frac{7}{3}x^2 + \frac{4}{3}x^2$

f) $12x^5 - x^5 - 4x^5 - 2x^5 - 3x^5$

g) $\frac{7}{4}x^5 - \frac{1}{4}x^5$

h) $x^2y^2 - 5x^2y^2 - (3x^2y^2 - 4x^2y^2) - 8x^2y^2$

i) $x^2 + \frac{x^2}{3}$

j) $x^2 + x^2$

k) $\frac{1}{2}x^3 - \frac{5}{2}x^3 + \frac{3}{2}x^3$

l) $-(ab^3 + a^3b) - 3a^3b + 5ab^3 - (a^3b - 2ab^3)$

m) $7x^3 - \frac{1}{2}x^3 - \frac{5}{2}x^3 + 2x^3 + \frac{3}{2}x^3$

n) $-x + x^2 + x^3 + 3x^2 - 2x^3 + 2x + 3x^2$

Exercises

Product and quotient of two monomials

To **multiply** two or more monomials, we must multiply, <u>separately</u>, their coefficients and literal parts.

Remember the method to multiply powers having the same base:
$$x^m \cdot x^n = x^{(m+n)}.$$

Example: Multiply: a) $5x^2 \cdot 7x^3$ b) $\dfrac{3}{2}x^2 \cdot 4x^3$

Solution:

a) $5x^2 \cdot 7x^3 = \left\{ \begin{array}{l} 5 \cdot 7 = 35 \\ x^2 \cdot x^3 = x^{(2+3)} = x^5 \end{array} \right\} = 35x^5$

b) $\dfrac{3}{2}x^2 \cdot 4x^3 = \left\{ \begin{array}{l} \dfrac{3}{2} \cdot 4 = \dfrac{3}{2} \cdot \dfrac{4}{1} = \dfrac{12}{2} = 6 \\ x^2 \cdot x^3 = x^{(2+3)} = x^5 \end{array} \right\} = 6x^5$

To **divide** two or more monomials, we must divide, <u>separately</u>, their coefficients and literal parts.

Remember the method to divide powers having the same base:

$$x^m : x^n = x^{(m-n)}.$$

Example: Divide: a) $15x^6 : 3x^2$ b) $\dfrac{3}{2}x^7 : \dfrac{9}{8}x^3$

Solution:

a) $15x^6 : 3x^2 = \left\{ \begin{array}{l} 15 : 3 = 5 \\ x^6 : x^2 = x^{(6-2)} = x^4 \end{array} \right\} = 5x^4$

b) $\dfrac{3}{2}x^7 : \dfrac{9}{8}x^3 = \left\{ \begin{array}{l} \dfrac{3}{2} : \dfrac{9}{8} = \dfrac{3}{2} \cdot \dfrac{8}{9} = \dfrac{3 \cdot 2 \cdot 2 \cdot 2}{2 \cdot 3 \cdot 3} = \dfrac{2 \cdot 2}{2} = \dfrac{4}{3} \\ x^7 : x^3 = x^{(7-3)} = x^4 \end{array} \right\} = \dfrac{4}{3}x^4$

Exercises

2. Calculate:

a) $7x \cdot (-8x^2)$

b) $(-3y^2) \cdot (-2y^3)$

c) $3x^2y \cdot 6xy^3$

d) $\dfrac{3}{4}x^2 \cdot \dfrac{5}{2}x^3$

e) $4a^3b^2 \cdot a^2b \cdot 7ab$

f) $-\dfrac{1}{2}a^3 \cdot \dfrac{5}{3}a^4$

g) $2a^6 \cdot 3a^6 \cdot 2a^6$

h) $\dfrac{2}{5}x^3 \cdot \left(-\dfrac{3}{2}x\right)$

i) $ab^3 \cdot (-3a^2b) \cdot 5a^3b$

j) $x^2 \cdot \dfrac{1}{3}x^5$

k) $-ab^2c^3 \cdot (-3a^2bc) \cdot 3abc$

l) $(6x^4) : (2x^2)$

m) $\dfrac{12a^6}{3a^3}$

n) $15x^4 : (-3x)$

ñ) $\dfrac{-14x^7}{7x^2}$

o) $-8x^4 : (-4x^3)$

p) $\dfrac{5x^7 y^3}{x^2 y}$

q) $(-18x^4) : (6x^3)$

r) $\dfrac{-12a^5 b^4 c^6}{2a^3 b^2 c}$

s) $2x^4 \cdot 6x^3 : 4x^2$

3. Work out the sum of the following monomials:

a) $5x + 3x^2 - 11x + 8x - x^2 + 7x$

b) $6x^2y - 13x^2y + 3x^2y - x^2y$

c) $2x - 5x^2 + 3x + 11y + 2x^3$

d) $3yz^3 + y^3z - 2z^3y + 5zy^3$

4. Work out the following products of monomials:

a) $(3x) \cdot (5x^2)$

b) $(-3x^2) \cdot (4x^3)$

c) $\left(\dfrac{2}{3}x^3\right) \cdot (-6x)$

d) $\left(\dfrac{2}{9}x^2\right) \cdot \left(-\dfrac{3}{5}x^3\right)$

e) $(7xy^2) \cdot (2y)$

f) $(5xyz) \cdot (-3x^2z)$

5. Work out the following products of monomials:

a) $(-2x^3) \cdot \left(\dfrac{4}{5}x^2\right) \cdot \left(\dfrac{1}{2}x\right)$

b) $\left(-\dfrac{5}{7}x^7\right) \cdot \left(\dfrac{3}{5}x^2\right) \cdot \left(-\dfrac{4}{3}x\right)$

c) $5x^3 \cdot 3x^2y \cdot (-4xz^3)$

d) $-3ab^2 \cdot 2ab \cdot (-\dfrac{2}{3}a^2b)$

e) $(3x^4 - 2x^3 + 2x^2 + 5) \cdot 2x^2$

f) $(-2x^5 + 3x^3 - 2x^2 - 7x + 1) \cdot (-3x^3)$

g) $\left(\dfrac{2}{3}x^3 - \dfrac{3}{2}x^2 + \dfrac{4}{5}x - \dfrac{5}{4}\right) \cdot 12x^2$

h) $\left(\dfrac{1}{2}ab^3 - a^2 + \dfrac{4}{3}a^2b + 2ab\right) \cdot 6a^2b$

6. Work out:

a) $8x^4 + 5x^4$

b) $7x^3 - 9x^3$

c) $-2x^2 + 4x^5 + 3x^2 - x^5$

d) $4x^7 - 2x + 5x - 4$

e) $3x^4 \cdot (-2x^3)$

f) $(-3x^8) \bullet (-6x^3)$

7. Work out the following combined operations with monomials:

a) $15x^5 - 3x^3 \cdot 4x^2$

b) $2x^3 + 4x^3 \cdot 5x - 2x \cdot (-x^2)$

c) $3a \cdot ab - 2a^2 \cdot (-4b) - 8 \cdot (2a^2b)$

d) $3x^2 + 4x^2 - 2x^2 \cdot (-3x) - [(4x^3 + x^2 - 2x \cdot (x^2)]$

e) $-3xy^2 - (-4x \cdot 7y^2) + (8x^2y^3 : 2xy)$

f) $(-y^2) \cdot (-2y^2) - 5y \cdot (-2y^3) + 3y^3 \cdot (-4y)$

g) $(3x^3 \cdot 6x - 2x^2 \cdot x^2) : (4x^2 \cdot 3x^2 - 8x \cdot x^3)$

h) $3x^5 - \dfrac{4}{3}x^2 \cdot \dfrac{3}{2}x^3$

i) $4a^2b \cdot (-ab^2) \cdot 5ab - 8a^4b^4$

j) $a^5 + \dfrac{5}{6}a^3 \cdot \dfrac{3}{5}6^2$

k) $5x^6 - 2x^6 \cdot 3x^6 : (-2x^6)$

l) $\left(-\dfrac{7}{3}x^3\right) \cdot \left(-\dfrac{4}{7}x\right) + \dfrac{2}{3}x^4$

m) $2ab \cdot (-a^3b) + [ab^2 \cdot (-3a^2b)] - 5a^3b \cdot ab + ab \cdot a^2b^2$

36

2. Polynomials. Addition, subtraction and product

Remember we cannon add or subtract unlike monomials. We must leave these operations indicated. These sums or subtractions of monomials are named polynomials.

For example, we have the polynomial $9x^3 + 8x^2 + 6x + 12$.

A *polynomial* in the unknown x is an algebraic expression given by the sum or subtraction of two or more monomials in the same unknown.

- We name *term* to each of the monomials in the polynomial. The monomial with degree 0 is named *independent term*. So, *binomial* is a polynomial with two monomials, e.g. 3x +5; *trinomial* is a polynomial with three monomials, e.g. $3x^2 + 5x + 2$.

- We name *degree* of a polynomial to the highest of the degrees of the monomials in it.

- A polynomial is *complete* when it has all the terms, from the term with degree 0 to the term with the highest degree.

- A polynomial is *ordinated* when the degrees of the terms are increasing or decreasing. Usually, monomials are ordinated in a decreasing order.

Polynomials are named as $A(x)$, $B(x)$, etc. indicating in the parentheses the unknown. So, the following expressions are polynomials.

Polynomial	Number of terms	Independent term	Degree	Complete	Ordinated
$A(x) = 5x^4 - 3x^2 + x - 1$	4	-1	4	Not	Yes
$B(x) = 2 - 5x^2 + 4x$	3 (trinomial)	2	2	Yes	Not
$C(x)$ 3x - 7	2 (binomial)	-7	1	Yes	Yes

2.1. Sum and subtraction of polynomials

Adding or subtracting polynomials consists simply in adding or subtracting their like monomials. Given two polynomials $A(x) = x^3 - 2x^2 - 7x - 4$ and $B(x) = -x^4 + 8x^2 + 7x + 2$, we can calculate A(x) + B(x) by writing one below the other, so that we collocate in the same column the like monomials. After that, we add the like monomials.

$$A(x) = \qquad x^3 - 2x^2 - 7x - 4$$
$$B(x) = -x^4 \qquad + 8x^2 + 7x + 2$$
$$\overline{A(x) + B(x) = -x^4 + x^3 + 6x^2 \qquad - 2}$$

Opposite polynomial of a given polynomial is another one that has the same monomials but with opposite coefficients (opposite signs).

If $B(x) = -x^4 + 8x^2 + 7x + 2$, its opposite polynomial is $-B(x) = x^4 - 8x^2 - 7x - 2$.

Do you agree that $3 - 2 = 3 + (-2)$?   So, in the same way, we can calculate $A(x) - B(x) = A(x) + [-B(x)]$:

$$
\begin{aligned}
A(x) = \quad & x^3 - 2x^2 - 7x - 4 \\
-B(x) = x^4 \quad & - 8x^2 - 7x - 2 \\
\hline
A(x) - B(x) = x^4 + & x^3 - 10x^2 - 14x - 6
\end{aligned}
$$

Another method to add or subtract polynomials is the application of the sign's rule for addition or subtraction of like monomials.

First of all, we remove the parentheses:
- If it is preceded by a + sign, we conserve the original signs.
- If it is preceded by a – sign, we change the sign inside the parenthesis.

We are using this method to calculate again $A(x) + B(x)$ and $A(x) - B(x)$:

$A(x) + B(x) = (x^3 - 2x^2 - 7x - 4) + (-x^4 + 8x^2 + 7x + 2) = x^3 - 2x^2 - 7x - 4 - x^4 + 8x^2 + 7x + 2 = -x^4 + x^3 + 6x^2 - 2$

$A(x) - B(x) = (x^3 - 2x^2 - 7x - 4) - (-x^4 + 8x^2 + 7x + 2) = x^3 - 2x^2 - 7x - 4 + x^4 - 8x^2 - 7x - 2 = x^4 + x^3 - 10x^2 - 14x - 6$

Exercises

8. Given polynomials $P(x) = 2x^5 - 3x^4 + 3x^2 - 5$ and $Q(x) = x^5 + 6x^4 - 4x^3 - x + 7$, work out

$P(x) + Q(x)$ and $P(x) - Q(x)$

9. Given polynomials $P(x) = 4x^3 + 6x^2 - 2x + 3$, $Q(x) = 2x^3 - x + 7$ and $R(x) = 7x^2 - 2x + 1$, work out:
a) $P(x) + Q(x) + R(x)$ b) $P(x) - Q(x) - R(x)$ c) $P(x) + 3Q(x) - 2R(x)$

2.2. Product of two polynomials

To **multiply** two polynomials, we must multiply each monomial of the first one by all the monomials of the second one, or vice versa.

For example, given the polynomials $A(x) = x^3 + 2x^2 - x + 2$ y $B(x) = 4x - 2$, we calculate $A(x) \times B(x)$:

$$
\begin{aligned}
A(x) = \quad & x^3 + 2x^2 - x + 2 \\
\times \quad B(x) = \quad & 4x - 2 \\
\hline
& -2x^3 - 4x^2 + 2x - 4 \\
4x^4 + 8x^3 - 4x^2 + 8x & \\
\hline
A(x) \cdot B(x) = 4x^4 + 6x^3 - 8x^2 + 10x - 4 &
\end{aligned}
$$

As we have already said, we can multiply two polynomials by applying the distributive property. We mean:

$$A(x) \times B(x) = (x^3 + 2x^2 - x + 2) \times (4x - 2) = x^3(4x - 2) + 2x^2(4x - 2) - x(4x - 2) + 2(4x - 2) =$$
$$= 4x^4 - 2x^3 + 8x^3 - 4x^2 - 4x^2 + 2x + 8x - 4 = 4x^4 + 6x^3 - 8x^2 + 10x - 4$$

Exercises

10. Work out the products of the following polynomials:

a) $(3x^2+5x-6)(8x^2-3x+4)$

b) $(5x^3-4x^2+x-2)(x^3-7x^2+3)$

c) $(2x^4-3x^2+5x)(3x^5-2x^3+x-2)$

d) $(ab^2+a^2b+ab)(ab-ab^2)$

e) $(-x^6+x^5-2x^3+7)(x^2-x+1)$

f) $(x^2y^2-2xy)(2xy+4)$

g) $(x^2-4x+3/2)(x+2)$

3. Remarkable identities. Special products

When working on polynomials, there are three identities that are going to be very useful for you to save time. They are named *polynomial identities*, *remarkable identities* or *special products*. They are:

Polynomial or Remarkable identities	
Binomial squares	**Sum x Difference:**
$(a+b)^2 = a^2 + 2ab + b^2$	
$(a-b)^2 = a^2 - 2ab + b^2$	$(a+b)\cdot(a-b) = a^2 - b^2$

The following two figures may help you to memorize these identities. They also are useful as demonstrations.

$(a+b)^2 = a^2 + 2ab + b^2$ $(a-b)^2 = a^2 - 2ab + b^2$

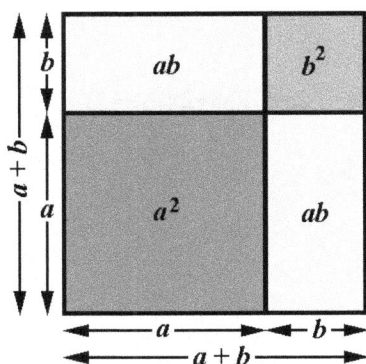

Squared of the sum of two monomials equals the squared of the first monomial plus double of the product of both monomials plus squared of the second one.

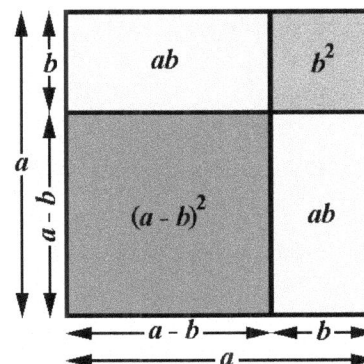

Squared of the difference of two monomials equals the squared of the first monomial minus double of the product of both monomials plus squared of the second one.

Example: Work out the following polynomial identities: a) $(x + 5)^2$ b) $(x - 6)^2$ c) $(x+2)\cdot(x-2)$

Solution:

a) $(x + 5)^2 = x^2 + 2\cdot x\cdot 5 + 5^2 = x^2 + 10x + 25$

b) $(3x - 4)^2 = (3x)^2 - 2\cdot 3x\cdot 4 + 4^2 = 9x^2 - 24x + 16$

c) $(x + 2)\cdot(x - 2) = x^2 - 2^2 = x^2 - 4$.

11. Work out the following products:

a) $(x + 2)^2$ b) $(x - 3)^2$ c) $(x + 4)(x - 4)$

d) $(x + 3)^2$ e) $(x - 4)^2$ f) $(x + 5)(x - 5)$

g) $(a + 4)^2$ h) $(a - 2)^2$ i) $(a + 3)(a - 3)$

j) $(2x + 3)^2$ k) $(3x - 2)^2$ l) $(2x + 1)(2x - 1)$

m) $(3x + 2)^2$ n) $(2x - 5)^2$ ñ) $(3x + 2)(3x - 2)$

o) $(4b + 2)^2$ p) $(5b - 3)^2$ q) $(b + 1)(b - 1)$

r) $(4a + 5)^2$ s) $(5a - 2)^2$ t) $(5a + 2)(5a - 2)$

u) $(4y + 1)^2$ v) $(2y - 3)^2$ w) $(2y + 3)(2y - 3)$

x) $(3x + 4)^2$ y) $(3x - 1)^2$ z) $(3x + 4)(3x - 4)$

a′) $(5b + 1)^2$ b′) $(2x - 4)^2$ c′) $(4x + 3)(4x - 3)$

12. Work out the following products:

a) $(2x^2+3x)^2$ b) $(2x^2-3)^2$ c) $(-x-3)^2$ d) $\left(x+\dfrac{1}{2}\right)^2$

e) $\left(2a-\dfrac{3}{2}\right)^2$ f) $\left(1+\dfrac{x}{2}\right)\left(1-\dfrac{x}{2}\right)$ g) $\left(2x+\dfrac{3}{4}\right)^2$ h) $\left(\dfrac{3}{2}-\dfrac{x}{4}\right)^2$

i) $\left(2+\dfrac{a}{3}\right)\left(-\dfrac{a}{3}+2\right)$ j) $\left(\dfrac{3x}{2}-\dfrac{1}{x}\right)^2$ k) $\left(\dfrac{x^2}{2}-\dfrac{x}{3}\right)\left(\dfrac{x^2}{2}+\dfrac{x}{3}\right)$ l) $\left(\dfrac{3}{2}x+\dfrac{1}{4}\right)^2$

13. Work out and simplify:

a) $(x+1)^2+(x-2)(x+2)$ b) $(3x-1)^2-(2x+5)(2x-5)$ c) $(2x+3)(-3+2x)-(x+1)^2$

d) $(-x+2)^2-(2x+1)^2-(x+1)(x-1)$ e) $-3x+x(2x-5)(2x+5)-(1-x^2)^2$

f) $(3x-1)^2-(-5x^2-3x)^2-(-x+2x^2)(2x^2+x)$

40

14. Work out and simplify:

a) $(2y + x)(2y - x) + (x + y)^2 - x(y + 3)$

b) $3x(x + y) - (x - y)^2 + (3x + y)y$

c) $(2y + x + 1)(x - 2y) - (x + 2y)(x - 2y)$

15. Work out and simplify:

a) $\dfrac{3x(x + 5)}{5} - \dfrac{(2x + 1)^2}{4} + \dfrac{(x - 4)(x + 4)}{2}$

b) $\dfrac{(8x^2 - 1)(x^2 + 2)}{10} - \dfrac{(3x^2 + 2)^2}{15} + \dfrac{(2x + 3)(2x - 3)}{6}$

4. Quotient of two polynomials

Quotient of two polynomials follows the same rule tan quotient of two whole numbers.

In whole numbers, given two numbers, named *dividend* and *divisor*, we find another two numbers, named *quotient* and *remainder*, such that: $D = d \times q + r$ con $r < d$:

The same in applied to polynomials:

Calculating the quotient of two polynomials, named ***dividend, D(x),*** and ***divisor, d(x),*** consists on finding another two polynomials, named **quotient, q(x),** and **remainder, r(x),** such that:

$$D(x) = d(x) \times q(x) + r(x); \qquad \text{degree } r(x) < \text{degree } d(x)$$

We are showing an example, the method to divide two polynomials,

$D(x) = 3x^3 - 2x^2 + 5x - 12$ and $d(x) = x^2 - 3x - 5$.

$$3x^3 - 2x^2 + 5x - 12 \quad \big|\underline{x^2 - 3x - 5}$$

Dividing two polynomials

Step 1: Write both polynomials in a decreasing order of their terms. If dividend is not a **complete** polynomial, leave gaps in the places of the non-existing terms.	$3x^3 - 2x^2 + 5x - 12 \ \lfloor \underline{x^2 - 3x - 5}$
Step 2: Calculate the quotient of the first term of dividend divided by the first term of the divisor. $3x^3 : x^2 = 3x$	$3x^3 - 2x^2 + 5x - 12 \ \lfloor \underline{x^2 - 3x - 5}$ $\qquad\qquad\qquad\qquad\quad 3x$
Step 3: obtained term of the quotient is multiplied by the whole divisor, and the resulting polynomial is subtracted to the dividend. In this way, we are obtaining the first *partial remainder*.	$3x^3 - 2x^2 + 5x - 12 \ \lfloor \underline{x^2 - 3x - 5}$ $\underline{-3x^3 + 9x^2 + 15x} \qquad\quad 3x$ $\qquad\quad 7x^2 + 20x$
Step 4: We write down the following term of the dividend after the remainder. We divide the first term of the first partial dividend by the first term of the divisor. We continue this process till obtaining a remainder with degree lower than the divisor´s degree.	$3x^3 - 2x^2 + 5x - 12 \ \lfloor \underline{x^2 - 3x - 5}$ $\underline{-3x^3 + 9x^2 + 15x} \qquad 3x + 7$ $\qquad\quad 7x^2 + 20x - 12$ $\qquad\quad \underline{-7x^2 + 21x + 35}$ $\qquad\qquad\qquad 41x + 23$

Notice that   degree q(x) = degree D(x) – degree d(x)   and it is a general rule.

16. Work out the quotient and the reminder of the following divisions:

a) $(7x^2 - 5x + 3) : (x^2 - 2x + 1)$ b) $(2x^3 - 7x^2 + 5x - 3) : (x^2 - 2x)$

c) $(x^3 - 5x^2 + 2x + 4) : (x^2 - x + 1)$ d) $(3x^5 - 2x^3 + 4x - 1) : (x^3 - 2x + 1)$

e) $(x^4 - 5x^3 + 3x - 2) : (x^2 + 1)$ f) $(4x^5 + 3x^3 - 2x) : (x^2 - x + 1)$

17. Divide and check that *Dividend = divisor x quotient + reminder*:

a) $(x^3 - 5x^2 + 3x + 1) : (x^2 - 5x + 1)$ b) $(6x^3 + 5x^2 - 9x) : (3x - 2)$

c) $(x^4 - 4x^2 + 12x - 9) : (x^2 - 2x + 3)$ d) $(4x^4 + 2x^3 - 2x^2 + 9x + 5) : (-2x^3 + x - 5)$

18. Divide:

a) $(x^4 - x^3 + 7x^2 + x + 15) : (x^2 + 2)$ b) $(2x^5 - x^3 + 2x^2 - 3x - 3) : (2x^2 - 3)$

c) $(6x^4 - 10x^3 + x^2 + 11x - 6) : (2x^2 - 4x + 3)$ d) $(x^3 + 2x^2 + x - 1) : (x^2 - 1)$

e) $(8x^5 - 16x^4 + 20x^3 - 11x^2 + 3x + 2) : (2x^2 - 3x + 2)$

Quotient of a polynomial by (x – a). Riffini´s rule

One of the most frequent situation when dividing polynomials is having as divisor a binomial like *(x – a)*, being *a* a real number. For example, $5x^4 - 4x^2 + 2x - 9) : (x - 4)$, where a = 4.
These division can be calculated in the general way, as above, but there is a more comfortable and quick method, named *Ruffinis´s Rule*. Riffini´s rule uses only the coefficients of the dividend.

We are going to show it with the example $x^4 - 8x^2 + 3x - 1 : (x - 2)$.

Ruffini's Rule

Step 1: Write the coefficients of dividend. You must list them in a decreasing order of degree. Write zero in the places of non-existing terms. Write the independent term of the divisor, <u>changing its sign</u>.	 1 0 −8 3 −1 2 _____
Step 2: Write down the first coefficient of the dividend.	$$\begin{array}{c|ccccc} & 1 & 0 & -8 & 3 & -1 \\ 2 & & & & & \\ \hline & 1 & & & & \end{array}$$
Step 3: Multiply the opposite of the independent term of the divisor, by this number you have just written down, and write it into the box, under the second term of the dividend.	$$\begin{array}{c|ccccc} & 1 & 0 & -8 & 3 & -1 \\ 2 & & 2 & & & \\ \hline x & 1 & & & & \end{array}$$
Step 4: Add in vertical.	$$\begin{array}{c|ccccc} & 1 & 0 & -8 & 3 & -1 \\ & & + & & & \\ 2 & & 2 & & & \\ \hline & 1 & 2 & & & \end{array}$$
Step 5: Repeat the product of 2 by the last number you wrote down in the quotient zone.	$$\begin{array}{c|ccccc} & 1 & 0 & -8 & 3 & -1 \\ 2 & & 2 & 4 & & \\ \hline x & 1 & 2 & & & \end{array}$$

Final steps: Continue this process till the end. Last obtained number in the quotient line is "remainder". It is usually written in a box.

$$\begin{array}{c|ccccc} & 1 & 0 & -8 & 3 & -1 \\ 2 & & 2 & 4 & -8 & -10 \\ \hline & 1 & 2 & -4 & -5 & \boxed{-11} \leftarrow \text{Remainder} \end{array}$$

Important note: Remember that, in general, degree q(x) = degree D(x) – degree d(x). In this case, as divisor (x – a) has degree = 1, degree q(x) = degree D(x) – 1.

Example: Determine the value of k that makes $4x^3 + 16x^2 + k$ divisible by $x + 3$.

Solution:

The remainder of the division $(4x^3 + 16x^2 + k) : (x + 3)$ must be zero. Applying the Ruffini´s rule, we have:

$$
\begin{array}{c|cccc}
 & 4 & 16 & 0 & k \\
-3 & & -12 & -12 & 36 \\
\hline
 & 4 & 4 & -12 & k + 36 = 0
\end{array}
$$

From the equation $k + 36 = 0$, we obtain that $k = (-36)$.

Root of a polynomial: We say a is a **root** of a polynomial P(x) if quotient $\dfrac{P(x)}{x - a}$ is an exact division (remainder = 0).

Example: Check if 2 is a root of the polynomial $P(x) = 3x^2 - 12x + 12$

Solution: We must divide $(3x^2 - 12x + 12) : (x - 2)$, and we do it by using the Ruffini´s rule:

$$
\begin{array}{c|ccc}
 & 3 & -12 & 12 \\
2 & & 6 & -12 \\
\hline
 & 3 & -6 & 0
\end{array}
$$

As remainder = 0, 2 is a root of the given polynomial.

Exercises

19. By uing Ruffini´s Rule, work out the quotient and the reminder of the following divisions:

a) $(5x^3 - 3x^2 + x - 2) : (x - 2)$

b) $(x^4 - 5x^3 + 7x + 3) : (x + 1)$

c) $(-x^3 + 4x) : (x - 3)$

d) $(x^4 - 3x^3 + 5) : (x + 2)$

Factor and reminder theorems

> The **Reminder** of the division of a polynomial $P(x)$ by a binomial $(x - a)$ is the value of **P(a)**.

20. By uing Ruffini´s Rule, work out the quotient and the reminder of $P(x) = 2x^3 - 5x^2 + 7x + 3$ when divided by the following divisors and check the Reminder´s theorem:

     a) $(x - 2)$ b) $(x + 3)$ c) x

21. By uing Ruffini´s Rule, work out the quotient and the reminder of $P(x) = x^4 - 3x^2 + 7$ when divided by the following divisors and check the Reminder´s theorem:

     a) $(x - 2)$ b) $(x + 3)$ c) x

22. Check which of the following numbers, 1, –1, 2, –2, 3, –3, are roots of the following polynomials:

     a) $P(x) = x^3 - 2x^2 - 5x + 6$ b) $Q(x) = x^3 - 3x^2 + x - 3$

23. Check if the following polynmials can be exactly divides by $(x - 3)$ or by $(x + 1)$.

     a) $P_1(x) = x^3 - 3x^2 + x - 3$ b) $P_2(x) = x^3 + 4x^2 - 11x - 30$

     c) $P_3(x) = x^4 - 7x^3 + 5x^2 - 13$

24. The polynomial $x^4 - 2x^3 - 23x^2 - 2x - 24$ is divisible by $(x - a)$ for two integer values of a. Find them out and calculate the quotient in both cases.

Factorization of a polynomial

> If a, b, c, … are the roots of a polynomial $P(x)$, the **factorization** of $P(x)$ is:
> $$P(x) = (x - a) \cdot (x - b) \cdot (x - c)……$$

Example: Factorize the polynomial $P(x) = x^2 + 4x - 5$

Solution: Using either Ruffini´s Rule or Reminder´s theorem, we must find its roots. They are $x = -5$, $x = 1$.

So, Factorization is $P(x) = x^2 + 4x - 5 = (x + 5)(x - 1)$

25. Factorize the following polynomials:

     a) $x^2 + 8x + 15$ b) $7x^2 - 21x - 280$ c) $3x^2 + 9x - 210$

26. By using the Reminder´s theorem, find out an integer root and, after that, factorize:

a) $2x^2 - 9x - 5$ b) $3x^2 - 2x - 5$ c) $4x^2 + 17x + 15$ d) $-x^2 + 17x - 72$

27. Extract a common factor and use the remarkable identities to factorize the polynomials:

a) $3x^3 - 12x$ b) $4x^3 - 24x^2 + 36x$ c) $45x^2 - 5x^4$

d) $x^4 + x^2 + 2x^3$ e) $x^6 - 16x^2$ f) $16x^4 - 9$

28. Complete the factorization of the following expressions:

a) $(x^2 - 25)(x^2 - 6x + 9)$ b) $(x^2 - 7x)(x^2 - 13x + 40)$

29. Factorize and indicate the roots:

a) $x^3 + 2x^2 - x - 2$ b) $3x^3 - 15x^2 + 12x$ c) $x^3 - 9x^2 + 15x - 7$ d) $x^4 - 13x^2 + 36$

5. Algebraic fractions

An algebraic fraction is a fraction whose numerator and denominator are algebraic expressions.

For example, $\dfrac{4x^4 + 2x^3 - 5x^2 + 2x - 8}{2x^2 + 5x - 6}$

30. Check if the following algebraic fractions are equivalent:

a) $\dfrac{x-4}{3x-12}$ y $\dfrac{1}{3}$ b) $\dfrac{x^2+x}{2x}$ y $\dfrac{x}{2}$

c) $\dfrac{x+y}{x^2-y^2}$ y $\dfrac{1}{x-y}$ d) $\dfrac{x}{x^2-x}$ y $\dfrac{2}{2x-2}$

31. Factorize nuerator and denominator and simplify:

a) $\dfrac{3x-3y}{x-y}$ b) $\dfrac{ax-a^2}{ay-a^2}$ c) $\dfrac{x^2-1}{2x+2}$ d) $\dfrac{x^2-2x+1}{x^2-1}$ e) $\dfrac{x^2-5x+6}{x^2-3x+2}$

f) $\dfrac{x^3-x}{x^2+x}$ g) $\dfrac{x^4-1}{(x^2+1)(x-1)}$ h) $\dfrac{a^4-b^4}{4a^2+4b^2}$ i) $\dfrac{x^3+y^3}{xy+(x-y)^2}$

j) $\dfrac{a^3-5a^2+6a}{a^2-7a+10}$ k) $\dfrac{7x-7y+7z}{14z-14y+14x}$ l) $\dfrac{(x-y)^2+4xy}{x^2-y^2}$

m) $\dfrac{ax+bx^2}{b^2x^2+2abx+a^2}$ n) $\dfrac{x^4-y^2}{y-x^2}$

32. Transform into a common denominator and work out:

a) $\dfrac{1}{2x} - \dfrac{1}{4x} + \dfrac{1}{x}$

b) $\dfrac{2}{x^2} - \dfrac{1}{3x} + \dfrac{1}{x}$

c) $\dfrac{1}{x-1} - \dfrac{1}{x}$

d) $\dfrac{2}{x-2} + \dfrac{2}{x+2}$

33. Work out:

a) $\dfrac{x}{2} + \dfrac{3}{x} - 1$

b) $\dfrac{2}{x^2} - \dfrac{x+1}{3x}$

c) $\dfrac{x}{x-3} - \dfrac{3}{x}$

d) $\dfrac{x-3}{x+1} - \dfrac{x}{x+3}$

34. Work out:

a) $\dfrac{x}{3} \cdot \dfrac{2x+1}{x-1}$

b) $\dfrac{2}{x-1} \cdot \dfrac{x}{x+1}$

c) $\dfrac{1}{x-1} : \dfrac{x+1}{3x}$

d) $\dfrac{2x}{2x-3} : \dfrac{x+1}{2x+3}$

35. Work out. Simplify as far as possible:

a) $\left(\dfrac{1}{x} : \dfrac{1}{x+1} \right) \cdot \dfrac{x}{2}$

b) $\left(\dfrac{2}{x} - \dfrac{2}{x+2} \right) : \dfrac{x-2}{x}$

Review exercises

1. Check if the polynomial $-x^4 + 3x^2 - 16x + 6$ is divisible by $(x - a)$ for any integer value of a.

2. Calculate the value of m so that makes the polynomial $P(x) = x^3 - mx^2 + 5x - 2$ divisible by $x + 1$.

3. The reminder of the following division is (-8). $(2x^4 + kx^3 - 7x + 6) : (x - 2)$. Which is the value of k?

4. Find out the value of m that makes the polynomial $mx^3 - 3x^2 + 5x + 9m$ divisible by $x + 2$.

5. Check if there is any divisibility relationship between the following pairs of polynomials:
   a) $P(x) = x^4 - 4x^2$ and $Q(x) = x^2 - 2x$
   b) $P(x) = x^2 - 10x + 25$ and $Q(x) = x^2 - 5x$
   c) $P(x) = x^3 + x^2 - 12x$ and $Q(x) = x - 3$

6. Find out the value of k that makes the polynomial $P(x) = x^3 - x^2 + x + k$ multiple of $Q(x) = x^2 + 1$.

7. Factorize and indicate the roots:
   a) $x^3 - 2x^2 - 2x - 3$ b) $2x^3 - 7x^2 - 19x + 60$ c) $x^3 - x - 6$ d) $4x^4 + 4x^3 - 3x^2 - 4x - 1$

8. Show that $(x - 2)$ is a factor of $P(x)$, where $P(x) = x^3 - 3x^2 - 10x + 24$, and find the other two factors.

9. (a) Show that $(x - 3)$ is a factor of $x^3 + x^2 - 8x - 12$ and find the other two factors.
   (b) Sketch the graph of $y = x^3 + x^2 - 8x - 12$.
   (c) Solve the inequality $x^3 + x^2 - 8x - 12 > 0$.

10. When $2x^3 - x^2 - 13x + k$ is divided by $x - 2$ the remainder is -20. Show that k = -6.

11. a) Multiply $x^2 + 3x - 5$ by $2x^2 - x + 4$.

b) Find values of p, q, r, s such that $(2x+3)(px^3 + qx^2 + rx + s) = 2x^4 + 13x^3 + 7x^2 + 18$ for all values of x.

c) Simplify $(x+1)(3x^2 + 2x - 5) - (x+1)(2x^2 + x + 1)$ as the product of three factors.

12. Let $P(x) = 2x^3 - 9x^2 + x - 12$.

a) Use the factor theorem to find a factor $(x - a)$ of $P(x)$, where a is an integer.

b) Find the other two factors.

c) Sketch the graph of $y = P(x)$, marking the coordinates of the points where it meets the x-axis and y axis.

d) Solve $P(x) > 0$.

13. Let $g(x) = x^3 + 3x^2 - 13x - 15$.

a) Show that $g(-5) = 0$ and $g(3) = 0$.

b) Hence factorise $g(x)$.

c) Sketch the graph of $y = g(x)$.

d) Hence write down the full set of values of x for which $g(x) > 0$.

14. a) Factorise $x^3 - 4x^2 + x + 6$, given that $(x - 2)$ is a factor.

b) Solve the inequality $x^3 - 4x^2 + x + 6 > 2(x-3)(x+1)$.

15. Find the quotient and the remainder when $x^3 + 2x^2 + x - 3$ is divided by $(x - 4)$.

[Hint: Use long division to divide $x^3 + 2x^2 + x - 3$ by $(x - 4)$].

16. Factorise $x^3 - 4x^2 + x + 6$, given that $(x - 2)$ is a factor.

Solve the inequality $x^3 - 4x^2 + x + 6 > 2(x-3)(x+1)$

17. Complete the following fractions to make them equivalent:

a) $\dfrac{x^2 - x}{x^2 - 1} = \dfrac{}{x+1}$

b) $\dfrac{x}{2x+1} = \dfrac{x^2}{}$

c) $\dfrac{x}{x-3} = \dfrac{}{x^2 - 9}$

d) $\dfrac{2}{x+2} = \dfrac{}{x^2 + 4x + 4}$

18. Work out and simplify:

a) $\left(\dfrac{3}{x} - \dfrac{x}{3}\right) : \left(\dfrac{1}{x} + \dfrac{1}{3}\right)$

b) $\dfrac{x+1}{(x-1)^2} \cdot \dfrac{x^2 - 1}{x}$

c) $\left[\left(x + \dfrac{1}{x}\right) : \left(x - \dfrac{1}{x}\right)\right] \cdot (x - 1)$

d) $\dfrac{2}{x} \cdot \left(\dfrac{1}{x} : \dfrac{1}{x-1}\right)$

19. Work out:

a) $\dfrac{x-2}{x^2} + \dfrac{x+2}{x^2 - x} - \dfrac{1}{x^2 - 1}$

b) $\dfrac{2x}{x^2 + x - 2} - \dfrac{5}{x+2} - \dfrac{x-4}{3x+6}$

c) $\dfrac{x+2}{2x+1} - \dfrac{2}{4x^2 - 1} + \dfrac{x+1}{2x}$

20. Work out and simplify:

a) $\left(1 - \dfrac{x-1}{x}\right)\dfrac{x^2}{x+3} - 1$

b) $\left(\dfrac{1}{x} - \dfrac{1}{x+3}\right) : \dfrac{3}{x^2}$

c) $4 - \dfrac{1}{2x-1}\left(\dfrac{2}{x} - \dfrac{1}{x^2}\right)$

21. Work out:

a) $\dfrac{x+1}{x-1} + \dfrac{3}{x+1} - \dfrac{x-2}{x^2-1}$

b) $\dfrac{x^2}{x^2-2x+1} + \dfrac{2x+3}{x-1} - 3$

c) $\dfrac{2x-3}{x^2-9} - \dfrac{x+1}{x-3} - \dfrac{x+2}{x+3}$

22. Translate into algebraic language by using only one unknown:

   a) The quotient of two consecutive pair numbers.   b) a number minus its inverse.

   c) The inverse of a number plus the invers of the doble of that number.

   d) The addition of the inverse of two consecutive numbers.

23. Give the expression of the volume and the areaof this hexaedral.

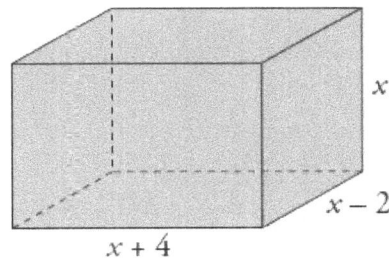

24. We are mixing x kg of an olive oil whose price is 5 €/kg and y kg of another oil whose price is 3 €/kg. Which is the price of the mixture? Express it asa function of x and y.

25. In a rectangle with sides x and y we inscribe a rhombus. Write the perimeter of the rhombus as a function of the sides of the rectangle.

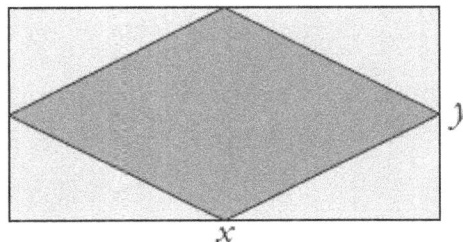

26. Give the algebraic expression of the shaded area, using only x and y.

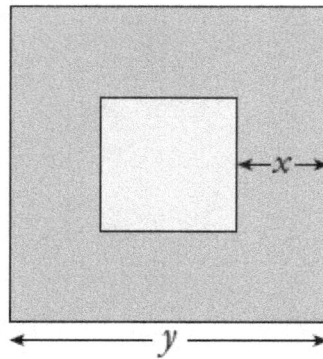

27. Two towns, A and B, are separated about 60 km. A car is leaving from A to B at a speed v. At the same time, another car is leavingfrom B to A at a speed v + 3. Find out the time they take to meet themselves, as a function of v.

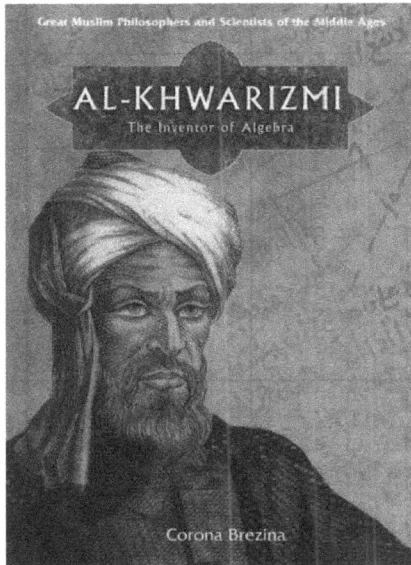

Great Muslim Philosophers and Scientists of the Middle Ages

AL-KHWARIZMI
The Inventor of Algebra

Corona Brezina

$$6x + 15 = 3x + 8$$
$$-3x \qquad -3x$$

ditch the smallest x guy

$$3x + 15 = 8$$
$$-15 \quad -15$$

ditch the 15

$$3x = -7$$

$$\frac{3x}{3} = \frac{-7}{3}$$

ditch the 3

$$x = \frac{-7}{3}$$

Unit 3.- Equations

1. First degree equations with one unknown

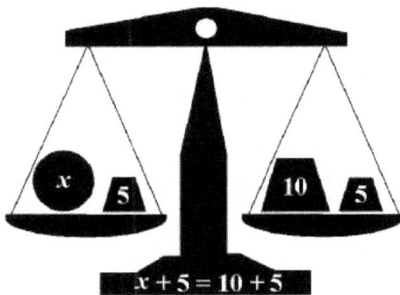

$$x + 5 = 10 + 5$$

As you can see in the figure, plates of the balance are equilibrated. So, there is an equality relationship between the masses of both plates.

We obtain an *equation*. Concretely, this one is named *first degree equation*. This means that the unknown has an exponent 1.

The following are also examples of first degree equations:

$$3x = 12 \qquad 3m + 7 = 2m - 6 \qquad 6t + 7 - 2t = 3t + 1$$

Solutions or *roots* of an equation are the values that, when substituted by the unknown, make the equality to verify.

Solving an equation is finding out the solutions or roots if the equation.

Depending on the solutions, an equation can be:

- *Compatible* if equation has solutions. If the number of solutions is finite, equation is named *determined compatible,* and if the number of solutions is infinite, equation is named *non-determined compatible.*

- *Incompatible* or *impossible* when equation has not solution.

1. Check if the given values are solutions of the following solutions:

a) $2x + 3 = 5x - 1 \rightarrow x = 3$

b) $\dfrac{x}{2} + 1 = x - 1 \rightarrow x = 4$

c) $3(x - 2) = 2x + 4 \rightarrow x = 10$

d) $\dfrac{5x - 2}{10} + 3 = 2x + 1 \rightarrow x = (-1)$

2. Write two equations having as solution x = (-2).

2. Equivalent equations. Equivalence transformations

Two or more equations are said to be *equivalent* if they have the same solution.

When we want to solve an equation, we must transform it into a more simple equation, equivalent to the original. This is possible by using the *Equivalence Transformations:*

Transformation		Practical rule
Adding or subtracting the same expression in both members of the equality.		What it is adding in a member, passes subtracting to the other member, and vice versa.
Multiplying or dividing both members of equality by the same number.		What it is multiplying in a member, passes dividing to the other member, and vice versa.

Example: By applying above instructions, solve these equations: a) $2x - 5 = x + 7$ b) $3x - 7 = 41$

Solution:

a) $2x - 5 = x + 7$
Add 5 to both members: $2x = x + 12$
Subtract x to both members $x = 12$
Solution is: $x = 12$

b) $3x - 7 = 41$
Add 7 to both members: $3x = 48$
Divide both members by 3: $x = 16$
Solution is: $x = 16$

3. Solve these equations:

a) 2x - 7 = 3x - 8   b) 2x - 4 = 3x + 2   c) 2x - 14 = -4x + 4

d) 18 + 3x = 2x - 12   e) 10 + 2x = 6x + 2   f) 9x - 9 = 6x + 18

g) 8 - 4 + x = 9   h) 12 - x + 3 = 6   i) 7 - 2x + 3 = 2

4. Apply equivalence transformations to solve the following equations:

a) $2x - 5 = x + 41 - 5$ b) $3x - 1 = x + 7$ c) $3x + 23 = 2x + 59$

d) $2x + \dfrac{5x}{3} = 66$ e) $5x + 8 = 8x + 2$ f) $9 + 9x = 117 - 3x$

g) $21 - 7x = 41x - 123$ h) $500 - 24x = -4 - 3x$

2.1. Steps to solve an equation

1.- Expand all **parentheses** in the equation.

2.- Eliminate **denominators** by multiplying every term by the LCM of the denominators.

3.- Move terms: Terms containing "x" must be together in one member, and non containing "x" terms, in the other one.

4.- Simplify **like terms** at both members, having an equation as **ax = b**.

5.- Calculate the value of the **unknown**.

Example: Solve the equation $\dfrac{x}{3} - \dfrac{3(x-2)}{5} = 2\left(\dfrac{x-1}{3}\right)$

Solution:

Step 1: Parentheses: $\dfrac{x}{3} - \dfrac{3x-6}{5} = \dfrac{2x-2}{3}$

Step 2: Denominators: LCM (3,5, 3) = 15. We multiply each term by 15:

$5x - 3(3x - 6) = 5(2x - 2)$ We expand new parenthesis: $5x - 9x + 18 = 10x - 10$

Step 3: Move terms: $5x - 9x - 10x = -10 - 18$

Step 4: Simplify like terms: $-14x = -28$

Step 5: and, finally, $x = \dfrac{-28}{-14} = 2$.

5. Solve:

a) $3(x+6) = -2(5-x)$

b) $x + 9 = 2(x-6)$

c) $2x + 3 = 4x + 6(x-4) - 2$

d) $1 + 4(x-2) = -3x + 5(x+1)$

e) $2(x+6) - 7x = 3x - 5x + 8$

f) $2(x+3) - 6(x+5) = 3x + 4$

g) $-2x + 3(x-1) = -12 + 5(2-x)$

h) $5(x-1) - 6x = 3(x-3)$

6. Solve:

a) $\dfrac{10x-55}{2} = 10x - \dfrac{95-10x}{2}$

b) $x + 5 = \dfrac{x+3}{3}$

c) $\dfrac{-4x+12}{4} = x - 5$

d) $\dfrac{3+x}{2} = 4$

e) $\dfrac{x}{2} + 21 = \dfrac{4x}{3} + 24$

f) $2(x+3) - 7 = 3\left(\dfrac{1}{2} - 2x\right)$

g) $\dfrac{5-9x}{8} + \dfrac{2x+3}{4} - \dfrac{143}{6} = 2x$

h) $\dfrac{5x+7}{2} - \dfrac{3x+9}{4} = \dfrac{2x+4}{3} + 5$

i) $1 - \dfrac{x-5}{4} - \dfrac{x-3}{10} + \dfrac{x+3}{8} = 0$

7. Solve these equations:

a) $\dfrac{3(x+2)}{2} + \dfrac{x-1}{5} = \dfrac{2(x+1)}{5} + \dfrac{37}{10}$

b) $\dfrac{2x-3}{2} - \dfrac{x+3}{4} = -4 - \dfrac{x-1}{2}$

c) $\dfrac{1+12x}{4} + \dfrac{x-4}{2} = \dfrac{3(x+1)-(1-x)}{8}$

d) $\dfrac{3x-2}{6} - \dfrac{4x+1}{10} = -\dfrac{2}{15} - \dfrac{2(x-3)}{4}$

e) $\dfrac{2x-3}{6} - \dfrac{3(x-1)}{4} - \dfrac{2(3-x)}{6} + \dfrac{5}{8} = 0$

f) $\dfrac{2}{3}(x+3) - \dfrac{1}{2}(x+1) = 1 - \dfrac{3}{4}(x+3)$

8. Solve these equations:

a) $3(x+6) = -2(5-x)$

b) $1 + 4(x-2) = -3x + 5(x+1)$

c) $2(x+3) - 6(x+5) = 3x + 4$

d) $3(5x+9) - 3(x-7) = 11(x-2)$

e) $\dfrac{10x-55}{2} = 10x - \dfrac{95-10x}{2}$

f) $\dfrac{x+4}{5} - \dfrac{x+3}{4} = 1 - \dfrac{x+1}{2}$

2.2. Number of solutions of a linear equation

As we have seen, all linear equations can be transformed into an equivalent one having the form **ax = b**. Depending of the values of a and b, the equation is going to a have a different number of solutions. We are going to see it with some examples:

- **180 - 3x = x + 12** $\rightarrow$ $-3x - x = 12 - 180$ $\rightarrow$ $-4x = -168$ $\rightarrow$ $x = 42$.

  This equation has an only solution.

- **7(x + 2) - 5x = 2x + 14** $\rightarrow$ $7x + 14 - 5x = 2x + 14$ $\rightarrow$ $7x - 5x - 2x = 14 - 14$ $\rightarrow$ $0x = 0$.

This equation has infinite solutions. All real numbers can be substituted by "x".

- $2(x + 2) = 2x + 7$ $\rightarrow$ $2x + 4 = 2x + 7$ $\rightarrow$ $2x - 2x = 7 - 4$ $\rightarrow$ $0x = 3.$

This equation has not solution. We cannot substitute any number by "x".

Remember:

Number of solutions of a linear equation

1°. Transform the equation into another equivalent one having the form $ax = b$.

2°. Pay attention to:

If $a \neq 0$, $ax = b$, equation has an *only solution*: x = b/a, so, it is a *determined compatible* equation.

If $a = b = 0$, $0x = 0$, equation has **infinite solutions** (all real numbers are solution), so, it is an *undetermined compatible* equation.

If $a = 0$ and $b \neq 0$, $0x = b$, equation has *no solution*, and it is an *incompatible* equation.

Exercises

9. Solve and classify these equations:

a) $3x + 100 = 5(200 - 3x)$

b) $5(20 - x) = 4(2x - 1)$

c) $-10x + 8 = 2(4 - 5x)$

d) $9(x - 2) - 3(x - 4) = 3(16 - 7x)$

e) $6 - (3x + 1) = 5 - 3(x - 2)$

f) $4 - 2(x - 1) = 3(2 - x) - 10$

g) $\dfrac{5x - 1}{7} - \dfrac{7x + 19}{20} = \dfrac{x + 11}{14} + \dfrac{3x - 4}{5}$

h) $\dfrac{x}{3} + 3 - \dfrac{2x}{5} = \dfrac{1}{5}\left(-\dfrac{x}{3} + 15\right)$

2.3. Problems with linear equations

Exercises

10. A pride of lions is made up of 13 lions, and there are 3 females more than males. How many lions and how many lionesses are there in the pride?

11. There are 31 people in a bar. How many men and women are there if we know that there are more men than women?

12. In a cow farm, the number of horns and legs add up to 30. How many cows are there in the farm?

13. When a number x is multiplied by 5, it gives the same result as when 48 is added to twice the number. Write down an equation for x and find its solution.

14. When 8 is added to a certain number, the result is 3 times as large as when 2 is subtracted from the number. Find the original number.

15. 60 people, between men, women and children, take part in a touristic trip. Children are double than men, and women number is 2/3 of the number of men and children together. How many men, women and children are there?

16. Find three consecutive multiples of 3 whose sum is 702.

17. Today, Mary´s age is double than Ann´s age. In 12 years, Mary´s age will be three times Ann´s age. How old are Mary and Ann today?

18. If we add 12 to certain number, calculate the double of the resulting quantity, and finally, subtract 24 units to the resulting number, we will obtain the double of the original number. What was the original number?

19. In a car park, there are 452 vehicles, between cars and motorbikes. Find out the number of vehicles of each kind, knowing there 1,744 wheels touching the floor.

20. In my school, we are 300 students (boys and girls). Today, 155 of them have had a cultural trip. Knowing they have taken part in the trip 60% of boys in the school and 40% of girls, how many boys and girls are there in the school?

21. A group of friends won a cash prize which gave them 28 € each. If there had been 5 more friends in the group they would each have received 7 € less. How many friends won the prize?

22. Janet is three times as old as her daughter, Mary. Five years ago, Janet was four times as old as Mary. How old is Janet now? How old will Mary be in 7 years?

3. Quadratic equations

It is named *quadratic equation* an equation that can be transformed into another equivalent equation having this form:

$$ax^2 + bx + c = 0, \quad \text{with } a \neq 0$$

Solutions of a quadratic equation are the values of x that being substituted in it, make the equation to verify.

Quadratic equations are solved by using this expression: $\quad x = \dfrac{-b \pm \sqrt{b^2 - (4 \cdot a \cdot c)}}{2 \cdot a}$

Example: Solve: $x^2 - 3x + 2 = 0$

Solution:

$$\left. \begin{array}{l} a = 1 \\ b = (-3) \\ c = 2 \end{array} \right\} x = \frac{-(-3) \pm \sqrt{(-3)^2 - (4 \cdot 1 \cdot 2)}}{2 \cdot 1} = \frac{+3 \pm \sqrt{9 - 8}}{2} = \frac{+3 \pm 1}{2} \quad \rightarrow \quad \left\{ \begin{array}{l} x_1 = \dfrac{+3 + 1}{2} = \dfrac{4}{2} = 2 \\ x_2 = \dfrac{+3 - 1}{2} = \dfrac{2}{2} = 1 \end{array} \right.$$

So, solutions are $\boxed{x = 2}$ and $\boxed{x = 1}$.

23. Solve these quadratic equations:

a) $x^2 - 6x + 8 = 0$ b) $x^2 - 8x + 7 = 0$ c) $4x^2 + 24x + 20 = 0$
d) $x^2 + 2x - 8 = 0$ e) $x^2 - 4x - 21 = 0$ f) $3x^2 + 6x - 12 = 0$

24. Solve these quadratic equations:

a) $2x^2 + 4x - 16 = 0$ b) $3x^2 - 6x - 9 = 0$ c) $x^2 + 9x + 20 = 0$
d) $x^2 + 4x + 4 = 16$ e) $x^2 - 6x + 9 = 25$ f) $4x^2 + 4x + 1 = 9$

25. Solve these equations:

a) $3 \cdot (x - 1) \cdot (x + 2) = 3x - 6$ b) $\dfrac{x}{5} \left(x + \dfrac{1}{6} \right) = x - 1$ c) $(5x - 3)^2 - 11(4x + 1) = 1$

d) $\dfrac{2x^2 - 1}{2} - \dfrac{x - 1}{3} = \dfrac{1 - x}{6}$ e) $(x - 1)(x + 1) + (x - 2)^2 = 3$ f) $\dfrac{x(x + 3)}{2} - \dfrac{(x + 1)^2}{3} + \dfrac{1}{3} = 0$

g) $(x + 2)(x - 3) + x = 3$ h) $(x + 1)^2 - 2x(x + 2) + 14 = 0$ i) $x(2x + 1) - \dfrac{(x - 1)^2}{2} = 3$

j) $(x + 1)^2 - (x - 1)^2 + 2 = x^2 + 6$ k) $(x + 2)^2 = 8x + 1$

3.2. Incomplete quadratic equations

In a quadratic equation $ax^2 + bx + c = 0$, if parameters b or c, are zero, it is named an *incomplete quadratic equation*.

Incomplete quadratic equations can be solved by using the general expression we have just studied, or by a shorter method:

- If $b = 0$, equation becomes $ax^2 + c = 0$. It is quickly solved.

- If $c = 0$, equation becomes $ax^2 + bx = 0$. It is solved by factorizing. **x(ax + b) = 0.**

Example: Solve: a) $3x^2 = 0$ b) $2x^2 - 8 = 0$ *c)* $x^2 - 12x = 0$

Solution:

   a) $3x^2 = 0$ Dividing by 3: $x^2 = 0 \rightarrow x = \sqrt{0} = \boxed{0}$.

   b) $2x^2 - 8 = 0$ Adding 8: $2x^2 = 8 \rightarrow$ Dividing by 2: $x^2 = 4 \rightarrow x = \sqrt{4} = \boxed{+2 \text{ and } (-2)}$.

   c) $x^2 - 12x = 0$ Factorizing: $x(x - 12) = 0 \rightarrow \begin{cases} x = 0 \\ x - 12 = 0 \rightarrow x = 12 \end{cases}$

For this last step, you must remember if you have a·b = 0, then, there are 2 possible solutions: $\begin{cases} a = 0 \\ b = 0 \end{cases}$

---

26. Solve these incomplete quadratic equations:
   a) $3x^2 - 12 = 0$ b) $5x^2 - 45 = 0$ c) $x^2 - 16 = 0$
   d) $2x^2 - 6x = 0$ e) $6x^2 - 18x = 0$

27. Solve these incomplete quadratic equations:
   a) $5x^2 - 20 = 0$ b) $3x^2 + 5x = 0$ c) $3x^2 - 75 = 0$ d) $5x^2 + 6x = 0$
   e) $3x^2 = 0$ f) $x^2 - 16 = 0$ g) $4x^2 + 7x = 0$ h) $5x^2 - 15x = 0$
   i) $2x^2 = 7x$ j) $3x^2 = 18x$

28. Solve these quadratic equations:
   a) $x^2 - 1 = 0$ b) $x^2 - 16 = 0$ c) $x(x + 1) = 0$ d) $x(x - 3) = 0$
   e) $(x - 1)^2 = 0$ f) $x^2 - 4 = 0$ g) $x^2 - 81 = 0$ h) $x(x - 1) = 0$
   i) $x(x + 5) = 0$ j) $(x - 3)^2 = 0$

29. Solve these incomplete quadratic equations, <u>without operating</u>:
   a) $(3x + 5) \cdot (2x - 1) = 0$ b) $(6x - 3) \cdot (2x - 1) = 0$ c) $(2x - 4) \cdot (x + 6) = 0$
   d) $(5x - 8) \cdot (x - 4) = 0$ e) $(x - 4)(x - 6) = 0$ f) $(x + 2)(x - 3) = 0$
   g) $x(x + 1)(x - 5) = 0$ h) $(3x + 1)(2x - 3) = 0$

30. Solve these quadratic equations:
   a) $x^2 + 2x = 0$ b) $2x^2 + 5x - 3 = 0$ c) $x^2 - 24 = 1$
   d) $1 - 4x^2 = -8$ e) $x^2 - 6x = 0$ f) $3x^2 - 39x = 0$

Exercises

3.3. Number of solutions of a quadratic equation

As you have already seen by solving all the previous quadratic equations, these equations can have:
- 2 solutions. This is the most general situation.
- 1 solution.
- No solution.

In the expression $x = \dfrac{-b \pm \sqrt{b^2 - (4 \cdot a \cdot c)}}{2 \cdot a}$, quantity into the square root is named ***discriminant*** (Δ).

So, discriminant is $\Delta = b^2 - (4 \cdot a \cdot c)$

Value of the discriminant allows to know the number of solutions of a quadratic equation, without having to calculate them:

- If $D > 0$ equation has **two solutions**.
- If $D = 0$ equation has **one solution** (named *double solution*).
- If $D < 0$ equation has **not solution**.

Example : Find the value of k so that equation $x^2 + kx + 36 = 0$ has two identical solutions.

Solution:

If we want the equation to have only one (double) solution, we need $D = 0$.

Substituting: $b^2 - 4ac = k^2 - 4 \cdot 1 \cdot 36 = k^2 - 144 = 0 \rightarrow k^2 = 144 \rightarrow k = \sqrt{144} = \pm 12$

So, equations are: $x^2 + 12x + 36 = 0$ and $x^2 - 12x + 36 = 0$.

---

31. Solve the following quadratic equations:
   a) $3x^2 - 5x + 2 = 0$ b) $16x^2 + 8x + 1 = 0$ c) $2x(x + 3) = 3(x - 19)$
   d) $2x^2 + 4x + 4 = 0$ e) $x^2 - 7x = 18$ f) $x^2 + 9 = 10x$
   g) $2x(3x - 4) + 2 = (1 - 3x)(x + 1)$ h) $4x^2 - 17x = -4$
   i) $4x^2 - 37x + 9 = 0$ j) $x^2 + x + 1 = 0$

32. Solve the following quadratic equations:
   a) $3x^2 - 5x + 2 = 0$ b) $16x^2 + 8x + 1 = 0$ c) $2x(x + 3) = 3(x - 19)$
   d) $2x^2 + 4x + 4 = 0$ e) $x^2 - 7x = 18$ f) $x^2 + 9 = 10x$
   g) $2x(3x - 4) + 2 = (1 - 3x)(x + 1)$ h) $4x^2 - 17x = -4$
   i) $4x^2 - 37x + 9 = 0$ j) $x^2 + x + 1 = 0$

33. In equation $x^2 - 5x + c = 0$, one solution is $x = 3$. a) What is the value of coefficient c? b) What is the other solution?

Exercises

34. Find out the value of m in equation $x^2 - 6x + m = 0$, so that it has two identical solutions. Find also the solutions.

35. Calculate the discriminant of each equation and, without solving it, indicate their number of solutions:
a) $5x^2 - 3x + 1 = 0$ b) $x^2 - 4x + 4 = 0$ c) $3x^2 - 6x - 1 = 0$
d) $5x^2 + 3x + 1 = 0$

36. Determine for what values of **m** equation $2x^2 - 5x + m = 0$:
a) Has two different solutions.   b) Has one solution.   c) Has not solution.

37. Decide for what values of **b** equation $x^2 - bx + 25 = 0$:
a) Has two different solutions.   b) Has one solution.   c) Has not solution.

38. Solve the following equations:

a) $\dfrac{1}{x} + \dfrac{1}{x^2} = \dfrac{3}{4}$ b) $\dfrac{2}{x} + \dfrac{3}{x^2} = 1$ c) $\dfrac{1}{x} + \dfrac{2}{x-1} = \dfrac{4}{3}$ d) $\dfrac{1}{x} - \dfrac{1}{x+3} = \dfrac{3}{10}$

e) $\dfrac{5}{x+2} + \dfrac{x}{x+3} = \dfrac{3}{2}$ f) $\dfrac{x-3}{x} + \dfrac{x+3}{x^2} = \dfrac{2}{3}$ g) $\dfrac{(x-2)^2}{x^2} - \dfrac{1}{2x} = \dfrac{8+3x}{2x^2} - \dfrac{2}{x}$

h) $\dfrac{3x+1}{x^3} + \dfrac{x+1}{x} = 1 + \dfrac{2x+3}{x^2}$ i) $\dfrac{x+1}{x-1} + \dfrac{3}{x+1} = \dfrac{x-2}{x^2-1}$

39. Solve these equations, by previous factorization:

a) $3x^3 - 12x = 0$ b) $4x^3 - 24x^2 + 36x = 0$ c) $45x^2 - 5x^4 = 0$

d) $x^4 + x^2 + 2x^3 = 0$ e) $x^6 - 16x^2 = 0$ f) $16x^4 - 9 = 0$

40. Solve these equations, by previous factorization:

a) $(x^2 - 25)(x^2 - 6x + 9) = 0$ b) $(x^2 - 7x)(x^2 - 13x + 40) = 0$

41. Solve these equations, by previous factorization:

a) $x^3 + 2x^2 - x - 2 = 0$ b) $3x^3 - 15x^2 + 12x = 0$

c) $x^3 - 9x^2 + 15x - 7 = 0$ d) $x^4 - 13x^2 + 36 = 0$

3.4. Problems with quadratic equations

42. The area of a parallelogram is 50 cm². If the base is twice its height, calculate the height.

43. The width of a rectangular plot of land is 5 m less than its length. If the area of the plot is 104 m², find the dimensions of the plot.

44. A circle has an area of 154 cm². Find its radius.

45. In a triangle, its base is 3 cm less than its height. If its area is 14 cm^2, find its height.

46. The area of a rectangle is 52 cm^2. Find the length and the width of this rectangle if their difference is 14 cm.

47. When 2 is added to a certain number, the result is the same as dividing 8 by the number. Solve this problem finding the number/s.

4. Biquadratic equations

Biquadratic equations are those that only have terms in x^4, x^2 and independent term.

For example, $3x^4 + 5x^2 - 2$

These equations are solved by means of the change: $\boxed{y = x^2}$

This change is shown in the following example.

Example: Solve the biquadratic equation $x^4 - 5x^2 + 4 = 0$

Solution: First of all, we change $\mathbf{y = x^2} \rightarrow y^2 - 5y + 4 = 0$

$$\left.\begin{array}{l} a = 1 \\ b = (-5) \\ c = 4 \end{array}\right\} y = \frac{-(-5) \pm \sqrt{(-5)^2 - (4 \cdot 1 \cdot 4)}}{2 \cdot 1} = \frac{+5 \pm \sqrt{25 - 16}}{2} = \frac{+5 \pm 3}{2} \rightarrow \left\{\begin{array}{l} y_1 = \frac{+5+3}{2} = \frac{8}{2} = 4 \\ y_2 = \frac{+5-3}{2} = \frac{2}{2} = 1 \end{array}\right.$$

So, solutions for "y" are y = 4 and y = 1. Now, we must undo the change we have introduced: $\mathbf{y = x^2}$

$y_1 = 4 \Rightarrow x^2 = 4 \Rightarrow x = \pm 2$

$y_2 = 1 \Rightarrow x^2 = 1 \Rightarrow x = \pm 1$

A biquadratic equation has four solutions, which can be repeated or not.

In our case, solutions are: $\boxed{x_1 = 2; \ x_2 = (-2); \ x_3 = 1; \ x_4 = (-1)}$.

48. Solve the following biquadratic equations:

a) $x^4 - 10x^2 + 9 = 0$ b) $x^4 - 2x^2 - 8 = 0$ c) $4x^4 + 7x^2 - 36 = 0$ d) $x^4 - 5x^2 + 4 = 0$

e) $x^4 - 13x^2 + 36 = 0$ f) $x^4 - 8x^2 - 9 = 0$ g) $\dfrac{2}{x^2 - 9} = \dfrac{x^2 - 16}{72}$ h) $x^4 - 5x^2 + 4 = 0$

i) $x^4 - 10x^2 + 9 = 0$ j) $x^4 + 3x^2 - 4 = 0$ k) $x^4 + 5x^2 + 4 = 0$ l) $x^4 - 16x^2 = 0$

m) $x^4 - 5x^2 - 36 = 0$ n) $x^4 - 5x^2 + 4 = 0$ ñ) $36x^4 - 13x^2 + 1 = 0$

49. Solve the following biquadratic equations:

a) $x^4 - x^2 = 0$ b) $x^4 + 20x^2 - 576 = 0$ c) $\dfrac{3 \cdot (x^2 - 11)}{5} - \dfrac{2 \cdot (x^2 - 60)}{7} = 36$

d) $x^4 - 26x^2 + 25 = 0$ e) $4x^4 - 37x^2 + 9 = 0$ f) $x^4 - 10x^2 + 9 = 0$

g) $4x^4 - 17x^2 + 4 = 0$ h) $x^4 - 4x^2 + 3 = 0$ i) $x^4 - 4x^2 = 0$

j) $2x^4 - 5x^2 + 2 = 0$ k) $x^4 + x^2 - 2 = 0$ l) $3x^4 - 2x^2 - 1 = 0$

m) $x^4 - 7x^2 + 12 = 0$

5. Equations with radicals

Some equations include radicals, usually square roots. In this section, we are learning to solve them.

OBJECTIVE: Isolate the radical in one of the members of the equation.

Once you have got it, you only have to calculate the squares of both members.

Example: Solve this equation: $\sqrt{2x - 3} - x = -1$

Solution: First of all, we isolate the radical in the first member → $\sqrt{2x - 3} = x - 1$

Now, we only have to calculate the squares of both members: $\left(\sqrt{2x - 3}\right)^2 = (x - 1)^2$

$2x - 3 = x^2 - 2x + 1$ → $x^2 - 4x + 4 = 0$ → Now, we only have tosolve this quadratic equation.

Their solutions are: $\boxed{x = 2 \text{ (double)}}$.

Sometimes, it is easier:

Example: Solve this equation: $2\sqrt{x + 4} = \sqrt{5x + 4}$

Solution: In this case, we can directly calculate the squares of both members: $\left(2\sqrt{x + 4}\right)^2 = \left(\sqrt{5x + 4}\right)^2$

$4(x + 4) = 5x + 4$ → $4x + 16 = 5x + 4$ →→→→ Its solution is: $\boxed{x = 12}$.

But, sometimes it is more difficult because you have to do it twice:

Example: Solve this equation: $\sqrt{2x+1}+\sqrt{x+4}=6$

<u>Solution:</u> First of all, we move one root to the right member: $\sqrt{2x+1}=6-\sqrt{x+4}$

Now, we calculate the squares of both members: $\left(\sqrt{2x+1}\right)^2=\left(6-\sqrt{x+4}\right)^2$

$2x+1=36+x+4-12\sqrt{x+4}$ → Now, we can isolate the radical at the left member: $12\sqrt{x+4}=-x+39$

Again, we calculate the squares of both members: $144(x+4)=x^2+1521+78x$ → $x^2-66x+945=0$

Now, we only have to solve this quadratic equation.

---

50. Solve the following equations with radicals:

a) $x+\sqrt{5x+10}=8$ b) $\sqrt{4x+5}-x=2$ c) $3\sqrt{6x+1}-5=2x$ d) $x-\sqrt{x}=2$

e) $\sqrt{x-9}=\sqrt{7-x}$ f) $x=2+\sqrt{2x-5}$ g) $\sqrt{x+1}+\sqrt{x-2}=3$ h) $x-\sqrt{3x-5}=3$

i) $\sqrt{6-3x}-\sqrt{3-x}=1$ j) $\sqrt{3x+1}-\sqrt{x+1}=\sqrt{x-4}$

51. Solve the following equations with radicals:

a) $\sqrt{x}-3=0$ b) $\sqrt{x}+2=x$ c) $\sqrt{4x+5}=x+2$ d) $\sqrt{x+1}-3=x-8$

e) $\sqrt{2x^2-2}=1-x$ f) $\sqrt{3x^2+4}=\sqrt{5x+6}$ g) $\sqrt{x^2+7}+2=2x$ h) $x-\sqrt{x}=2$

i) $x-\sqrt{25-x^2}=1$ j) $x-\sqrt{169-x^2}=17$ k) $x+\sqrt{5x+10}=8$ l) $\sqrt{x+4}=7$

m) $x+\sqrt{5x+10}=8$ n) $x-\sqrt{25-x^2}=1$

Exercises

---

6. Simultaneous equations. Systems of equations

Although you have already studied simultaneous equations, we are starting by reviewing the main methods to solve them, so that you get able to solve some more difficult cases, even systems of quadratic equations.

There are three methods to solve systems of linear equations:

- Substitution method

- Equalization method

- Reduction method

6.1. Substitution method

We are going to describe *substitution method* with an example.

Example: Solve the following system by using the substitution method.

$$\begin{cases} x - 2y = -2 \\ 2x + 2y = 8 \end{cases}$$

<u>Solution:</u> You must follow the following steps:

Step 1: Isolate one unknown, where easier.	⟹	In our case, easier unknown is "x" in the first equation: x = - 2 + 2y
Step 2: Substitute this expression in the <u>other</u> equation.	⟹	2x + 2y = 8. As x = -2+2y, then 2·(-2 + 2y) + 2y = 8
Step 3: Solve this new equation with only one unknown.	⟹	2·(-2 + 2y) + 2y = 8 - 4 + 4y + 2y = 8 4y + 2y = 4 + 8 6y = 12 $y = \dfrac{12}{6} = 2$ $\boxed{y = 2}$
Step 4: In order to obtain the value of the other unknown, we substitute known value in any of the two given equations. We choose it freely, but it is a good idea looking for the easiest one.	⟹	x = - 2 + 2y As y = 2, then x = - 2 + 2·2 = - 2 + 4 = $\boxed{2}$ So, solution is $\boxed{(x = 2,\ y = 2)}$.
Step 5: Check your solution.	⟹	$\begin{cases} x - 2y = -2 \rightarrow 2 - 2\cdot2 = 2 - 4 = -2 \rightarrow OK!!! \\ 2x + 2y = 8 \rightarrow 2\cdot2 + 2\cdot2 = 4 + 4 = 8 \rightarrow OK!!! \end{cases}$

Exercises

52. Solve the following systems by using the *substitution* method.

a) $\begin{cases} x + 3y = 5 \\ 5x + 7y = 13 \end{cases}$ b) $\begin{cases} 6x + 3y = 0 \\ 3x - y = 3 \end{cases}$ c) $\begin{cases} 3x + 9y = 4 \\ 2x + 3y = 1 \end{cases}$ d) $\begin{cases} x - 4y = 11 \\ 5x + 7y = 1 \end{cases}$

53. Solve the following systems by using the *substitution* method.

a) $\left.\begin{array}{l} 3x + 2y = 23 \\ x + y = 8 \end{array}\right\}$ b) $\left.\begin{array}{l} 3x + 2y = 14 \\ 2x - 5y = -16 \end{array}\right\}$ c) $\left.\begin{array}{l} x + 3y = 4 \\ 2x - y = 1 \end{array}\right\}$ d) $\left.\begin{array}{l} 3x - y = 17 \\ 2x + y = 8 \end{array}\right\}$

6.2. Equalization method

We are going to describe *equalization method* with an example.

Example: Solve the following system by using the equalization method: $\begin{cases} 3x - 2y = 1 \\ 2x + 5y = 7 \end{cases}$

Step 1: Isolate one unknown (choose freely) in both equations. As you are free, choose the easiest one.	➡	We are going to isolate the "x": - First equation: $3x - 2y = 1 \rightarrow 3x = 2y + 1 \rightarrow \boxed{x = \dfrac{2y+1}{3}}$ - Second equation: $2x + 5y = 7 \rightarrow 2x = -5y + 7 \rightarrow \boxed{x = \dfrac{-5y+7}{2}}$
Step 2: Build an equality with both expressions.	➡	$\dfrac{2y+1}{3} = \dfrac{-5y+7}{2}$
Step 3: Solve this new equation with only one unknown.	➡	LCM(3,2) = 6, we must mutiply all the terms by 6. $2(2y+1) = 3(-5y+7)$ $4y + 2 = -15y + 21$ $4y + 15 = -2 + 21$ $19y = 19 \rightarrow y = \dfrac{19}{19} = 1 \;\boxed{y = 1}$
Step 4: In order to obtain the value of the other unknown, we substitute known value in any of the two given equations. We choose it freely, but it is a good idea looking for the easiest one.	➡	$2x + 5y = 7 \quad$ As y = 1, then $2x + 5 \cdot 1 = 7$ $2x + 5 = 7$ $2x = 7 - 5 = 2$ $x = \dfrac{2}{2} = 1$ So, solution is $\boxed{(x = 1, \; y = 1)}$.
Step 5: Check your solution.	➡	$\begin{cases} 3x - 2y = 1 \rightarrow 3 \cdot 1 - 2 \cdot 1 = 3 - 2 = 1 \rightarrow OK!!! \\ 2x + 5y = 7 \rightarrow 2 \cdot 1 + 5 \cdot 1 = 2 + 5 = 7 \rightarrow OK!!! \end{cases}$

Exercises

54. Solve the following systems by using the *equalization* method.

a) $\begin{cases} x + 3y = 5 \\ 5x + 7y = 13 \end{cases}$
b) $\begin{cases} 6x + 3y = 0 \\ 3x - y = 3 \end{cases}$
c) $\begin{cases} 3x + 9y = 4 \\ 2x + 3y = 1 \end{cases}$
d) $\begin{cases} x - 4y = 11 \\ 5x + 7y = 1 \end{cases}$

55. Solve the following systems by using the *equalization* method.

a) $\left.\begin{array}{l} 7x - 5y = 52 \\ 2x + 5y = 47 \end{array}\right\}$
b) $\left.\begin{array}{l} 2x - y = 5 \\ x + 2y = 5 \end{array}\right\}$
c) $\left.\begin{array}{l} 3x - y = 7 \\ 2x + 3y = 1 \end{array}\right\}$
d) $\left.\begin{array}{l} 7x + 4y = 3 \\ 5x - 3y = 8 \end{array}\right\}$

6.3. Reduction method

Reduction method is based on the fact that there are operations that can be done on a system of equations without changing its solution. These operations are named ***equivalence operations***, as they transform a system into another different one, but equivalent to original one:

- Adding both equations.
- Multiplying one equation by a real number and, after, adding.
- Multiplying both equations by real numbers and, after, adding.

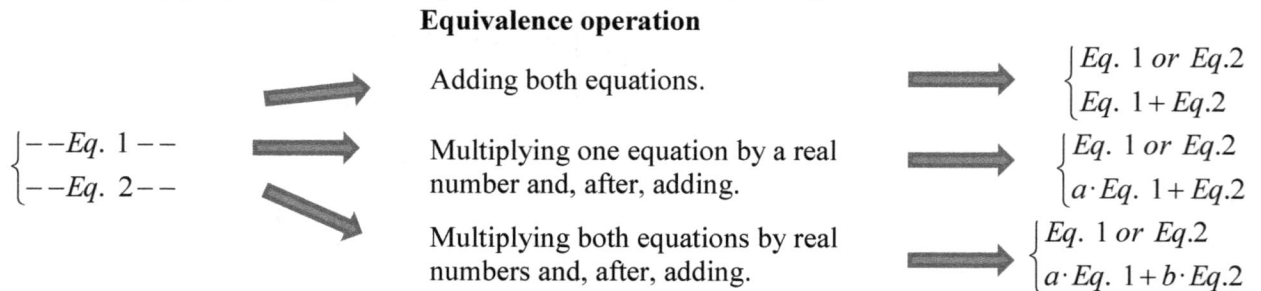

Equivalence operation

$$\begin{cases} --Eq.\ 1\ -- \\ --Eq.\ 2\ -- \end{cases}$$

Adding both equations. $\quad\longrightarrow\quad \begin{cases} Eq.\ 1\ or\ Eq.2 \\ Eq.\ 1+Eq.2 \end{cases}$

Multiplying one equation by a real number and, after, adding. $\quad\longrightarrow\quad \begin{cases} Eq.\ 1\ or\ Eq.2 \\ a\cdot Eq.\ 1+Eq.2 \end{cases}$

Multiplying both equations by real numbers and, after, adding. $\quad\longrightarrow\quad \begin{cases} Eq.\ 1\ or\ Eq.2 \\ a\cdot Eq.\ 1+b\cdot Eq.2 \end{cases}$

We are going to describe ***reduction method*** with examples.

Example: Solve the following systems by using the reduction method:

a) $\begin{cases} 3x-2y=1 \\ 5x+2y=7 \end{cases}$ b) $\begin{cases} 3x+y=1 \\ 5x+y=3 \end{cases}$ c) $\begin{cases} 3x+2y=1 \\ 5x+y=4 \end{cases}$ d) $\begin{cases} 3x+5y=1 \\ 2x+3y=1 \end{cases}$

<u>Solution:</u>

a) $\begin{cases} 3x-2y=1 \\ 5x+2y=7 \end{cases}$

Notice if you add both equations, terms containing unknown *"y"* will be cancellated.

$$\begin{cases} 3x-2y=1 \\ 5x+2y=7 \end{cases}$$

Addition: 8x $\qquad$ = 8 $\qquad$ Solving this equation is easy, being x = $\boxed{1}$.

In order to obtain the value of the other unknown, we substitute known value in any of the two given equations. We choose it freely, but it is a good idea looking for the easiest one.

$\qquad$ 5x + 2y = 7 $\quad$ As x = 1, then

$\qquad$ 5·1 + 2y = 7

$\qquad$ 5 + 2y = 7

$\qquad$ 2y = 7 − 5 = 2

$\qquad$ $x = \dfrac{2}{2} = 1$

$\qquad$ So, solution is $\boxed{(x = 1,\ y = 1)}$.

Finally, check your solution. $\begin{cases} 3x-2y=1 \to 3\cdot1-2\cdot1=3-2=1 \to OK!!! \\ 5x+2y=7 \to 5\cdot1+2\cdot1=5+2=7 \to OK!!! \end{cases}$

b) $\begin{cases} 3x+y=1 \\ 5x+y=3 \end{cases}$

Notice if you add both equations, none of the unknowns are being cancelled. We need a (-) sign before any of the *"y"* unknowns. So, we must multiply one of the equations by (-1).

$$\begin{cases} 3x + y = 1 \ \underline{Multiply \ by \ (-1)} \\ 5x + y = 3 \ \underline{Leave} \end{cases} \qquad \begin{cases} -3x - y = -1 \\ 5x + y = 3 \end{cases}$$

Now, if you add both equations, terms containing unknown *"y"* will be cancelled.

$$\begin{cases} -3x - y = -1 \\ 5x + y = 3 \end{cases}$$

Addition: $2x = 2$ 　　　　Solving this equation is easy, being x = $\boxed{1}$.

In order to obtain the value of the other unknown, we substitute known value in any of the two given equations. We choose it freely, but it is a good idea looking for the easiest one.

　　　$5x + y = 3$　As x = 1, then

　　$5 \cdot 1 + y = 3$

　　　$5 + y = 3$

　　　　$y = 3 - 5 = \boxed{(-2)}$

So, solution is $\boxed{(x = 1, \ y = (-2))}$.

Finally, check your solution.　$\begin{cases} 3x + y = 1 \rightarrow 3 \cdot 1 + (-2) = 3 - 2 = 1 \rightarrow OK!!! \\ 5x + y = 3 \rightarrow 5 \cdot 1 + (-2) = 5 - 2 = 3 \rightarrow OK!!! \end{cases}$

c) $\begin{cases} 3x + 2y = 1 \\ 5x + y = 4 \end{cases}$

Notice if you add both equations, none of the unknowns are being cancelled. We need a (-2) before *"y"* unknown in second equation. So, we must multiply second equation by (-2).

$$\begin{cases} 3x + 2y = 1 \ \underline{Leave} \\ 5x + y = 4 \ \underline{Multiply \ by \ (-2)} \end{cases} \qquad \begin{cases} 3x + 2y = 1 \\ -10x - 2y = -8 \end{cases}$$

Now, if you add both equations, terms containing unknown *"y"* will be cancelled.

$$\begin{cases} 3x + 2y = 1 \\ -10x - 2y = -8 \end{cases}$$

Addition: $-7x = -7$ 　　　　Solving this equation is easy, being x = $\boxed{1}$.

In order to obtain the value of the other unknown, we substitute known value in any of the two given equations. We choose it freely, but it is a good idea looking for the easiest one.

　　　$5x + y = 4$　As x = 1, then

　　$5 \cdot 1 + y = 4$　　$\rightarrow$　　$5 + y = 4$　$\rightarrow$　$y = 4 - 5 = \boxed{(-1)}$　　　So, solution is $\boxed{(x = 1, \ y = (-1))}$.

d) $\begin{cases} 3x+5y=1 \\ 2x+3y=1 \end{cases}$

Notice if you add both equations, none of the unknowns are being cancelled. Moreover, even multiplying any of the equations by a real number, we would not be able to cancel any unknown. When this happens, you must choose one of the unknowns and *cross-multiplying*.

For example, if we work on "x", we can multiply by 2 above and by (-3) below:

$$\begin{cases} 3x+5y=1 \; \underline{Multiply \; by \; 2} \\ 2x+3y=1 \; \underline{Multiply \; by \; (-3)} \end{cases} \longrightarrow \begin{cases} 6x+10y=2 \\ -6x-9y=-3 \end{cases}$$

Now, if you add both equations, terms containing unknown "x" will be cancelled.

$$\begin{cases} 6x+10y=2 \\ -6x-9y=-3 \end{cases}$$

Addition: y = - 1   So, y = $\boxed{(-1)}$.

In order to obtain the value of the other unknown, we substitute known value in any of the two given equations. We choose it freely, but it is a good idea looking for the easiest one.

2x + 3y = 1   As x = 1, then

2x + 3 · (-1) = 1

2x − 3 = 1

2x = 1 + 3 = 4 → 2x = 4 → x = $\boxed{2}$

So, solution is $\boxed{(x = 2, \; y = (-1))}$.

Finally, check your solution. $\begin{cases} 3x+5y=1 \rightarrow 3 \cdot 2 + 5 \cdot (-1) = 6 - 5 = 1 \rightarrow OK!!! \\ 2x+3y=1 \rightarrow 2 \cdot 2 + 3 \cdot (-1) = 4 - 3 = 1 \rightarrow OK!!! \end{cases}$

Exercises

56. Solve the following systems by using the *reduction* method.

a) $\begin{cases} 3x+5y=11 \\ 4x-5y=38 \end{cases}$ b) $\begin{cases} x+3y=5 \\ 5x+7y=13 \end{cases}$ c) $\begin{cases} x-4y=11 \\ 5x+7y=1 \end{cases}$ d) $\begin{cases} 6x+3y=0 \\ 3x-y=3 \end{cases}$

57. Solve the following systems by using the *reduction* method.

a) $\begin{cases} 7x-5y=52 \\ 2x+5y=47 \end{cases}$ b) $\begin{cases} 2x-y=5 \\ x+2y=5 \end{cases}$ c) $\begin{cases} 3x-y=7 \\ 2x+3y=1 \end{cases}$ d) $\begin{cases} 7x+4y=3 \\ 5x-3y=8 \end{cases}$

58. Solve the following systems by any method:

a) $\left.\begin{array}{l} x+y=2 \\ x-y=6 \end{array}\right\}$
b) $\left.\begin{array}{l} x+y=12 \\ x-y=2 \end{array}\right\}$
c) $\left.\begin{array}{l} x+y=5 \\ -x+y=1 \end{array}\right\}$
d) $\left.\begin{array}{l} 3x-4y=-6 \\ x+2y=8 \end{array}\right\}$

e) $\left.\begin{array}{l} 3x-y=17 \\ 2x+y=8 \end{array}\right\}$
d) $\left.\begin{array}{l} x+2y=7 \\ 7x-4y=13 \end{array}\right\}$

59. Solve the following systems:

a) $\left.\begin{array}{l} y-3x=-8 \\ 3y-5x=y-3 \end{array}\right\}$
b) $\left.\begin{array}{l} \dfrac{x}{3}+\dfrac{y}{5}=7 \\ \dfrac{x}{3}-\dfrac{y}{4}=-2 \end{array}\right\}$
c) $\left.\begin{array}{l} \dfrac{2x}{3}+\dfrac{3y}{4}=5 \\ \dfrac{5x}{3}-\dfrac{y}{4}=3 \end{array}\right\}$
d) $\left.\begin{array}{l} \dfrac{x}{2}+\dfrac{y}{4}=3 \\ x+2y=12 \end{array}\right\}$

60. Solve the following systems:

a) $\left.\begin{array}{l} x-(y+1)=3 \\ y+(x+3)=4 \end{array}\right\}$
b) $\left.\begin{array}{l} 10(x-2)+y=1 \\ x+3(x-y)=5 \end{array}\right\}$
c) $\left.\begin{array}{l} x+3=y-3 \\ 2(x+3)=6-y \end{array}\right\}$

d) $\left.\begin{array}{l} 3x+2=3(y+2)-4 \\ 5x-2y=5-x-y \end{array}\right\}$
e) $\left.\begin{array}{l} \dfrac{3(x-y)}{2}-\dfrac{2x+y}{3}=-3 \\ 2(x+y-1)+\dfrac{2x-y}{3}=11 \end{array}\right\}$
f) $\left.\begin{array}{l} \dfrac{x+y}{3}-\dfrac{2(x-3)}{5}=\dfrac{11}{5} \\ \dfrac{3x+y}{2}+2(x-y)=\dfrac{-9}{2} \end{array}\right\}$

g) $\left.\begin{array}{l} \dfrac{5(x+1)}{3}-\dfrac{x+y}{2}=\dfrac{7}{2} \\ 2(x-y+5)-\dfrac{x+y}{3}=11 \end{array}\right\}$
h) $\left.\begin{array}{l} \dfrac{2(x+1)}{5}-\dfrac{3(y-2)}{2}=0 \\ \dfrac{x+y}{4}=\dfrac{1}{4} \end{array}\right\}$

61. Solve the following systems:

a) $\left.\begin{array}{l} 2(x-4)-3(y-7)+22=0 \\ 2(x+1)+4(y+1)-16=0 \end{array}\right\}$
b) $\left.\begin{array}{l} x-7(y+4)=-5 \\ 2x-3y-19=-6 \end{array}\right\}$
c) $\left.\begin{array}{l} x-6(y-2)=-6 \\ 3(x-1)+2y=23 \end{array}\right\}$

d) $\left.\begin{array}{l} \dfrac{3x}{5}-\dfrac{2y}{3}=7 \\ \dfrac{5x}{3}-2y=2 \end{array}\right\}$
d) $\left.\begin{array}{l} \dfrac{x}{3}+\dfrac{y}{4}=x-\dfrac{1}{6} \\ \dfrac{y}{3}-\dfrac{x}{5}=\dfrac{x+y+4}{15} \end{array}\right\}$
e) $\left.\begin{array}{l} 2x-y=7 \\ \dfrac{4}{3}x-\dfrac{1}{8}y=\dfrac{19}{3}-4 \end{array}\right\}$

f) $\left.\begin{array}{l} 3y-2x-16=0 \\ 2(x-5)+6(y-2)+20=0 \end{array}\right\}$
g) $\left.\begin{array}{l} \dfrac{x-2}{3}+\dfrac{y-1}{4}-1=x \\ 4x+5y=-18 \end{array}\right\}$
h) $\left.\begin{array}{l} \dfrac{x-y}{2}-\dfrac{x+y}{10}=\dfrac{3}{5} \\ 3x-\dfrac{5y-4}{2}=\dfrac{25}{2} \end{array}\right\}$

62. Solve the following systems:

a) $\begin{aligned} 2x - y &= 7 \\ \dfrac{4x}{3} - \dfrac{y}{3} &= \dfrac{19}{3} - 4 \end{aligned}\Bigg\}$

b) $\begin{aligned} \dfrac{x}{3} + \dfrac{y}{4} &= x - \dfrac{1}{6} \\ \dfrac{y}{3} - \dfrac{x}{5} &= \dfrac{x+y+4}{15} \end{aligned}\Bigg\}$

c) $\begin{aligned} \dfrac{x-y}{2} - \dfrac{x+y}{10} &= \dfrac{3}{5} \\ 3x - \dfrac{5y-4}{2} &= \dfrac{25}{2} \end{aligned}\Bigg\}$

d) $\begin{aligned} \dfrac{x-y-1}{3} &= \dfrac{x}{4} - \dfrac{y}{5} + 1 \\ \dfrac{x}{2} + \dfrac{y-1}{3} - \dfrac{3}{4} &= \dfrac{x+y}{3} - \dfrac{x-y}{5} \end{aligned}\Bigg\}$

Systems of non-linear equations

Systems of non-linear equations are systems in which, at least one of the equations contain some quadratic terms.

For example, $\begin{aligned} x - y &= 1 \\ x^2 + y^2 - 2x &= 31 \end{aligned}\Bigg\}$

In this section, we are learning to solve these systems. Depending on the system, we can use substitution or reduction methods. We are seeing it with some examples.

Example: Solve the following system: $\begin{aligned} x - y &= 1 \\ x^2 + y^2 - 2x &= 31 \end{aligned}\Bigg\}$

Solution: When one of the equations contain quadratic terms but the other one does not, we will use substitution method: We can isolate "x" in the first equation: $x = y + 1$ and substitute it in the second one:

$(y+1)^2 + y^2 - 2(y+1) = 31 \Rightarrow y^2 + 2y + 1 + y^2 - 2y - 2 = 31 \Rightarrow 2y^2 = 32 \Rightarrow y^2 = 16 \Rightarrow \boxed{y \pm 4}$.

Now, we must find out the corresponding values for "x". We must substitute, for example, in the first one:

$y_1 = 4$;   Como $x - y = 1$; $x = y + 1 \rightarrow x_1 = 4 + 1 = 5$.

$y_2 = (-4)$; $x = y + 1 \rightarrow x_2 = -4 + 1 = (-3)$.   So, solutions are $\boxed{(5, 4) \text{ and } (-3, -4)}$.

Example: Solve the following system: $\left.\begin{array}{r} x^2 - 2y = 0 \\ 2x^2 + y = 0 \end{array}\right\}$

Solution: Notice if we multiply by (-2) the first equation and add both equations, x unknown will disappear.

$\left\{\begin{array}{l} x^2 - 2y = 0 \ \underline{Multiply \ by \ (-2)} \\ 2x^2 + y = 0 \ \underline{Leave} \end{array}\right.$ $\left\{\begin{array}{l} -2x^2 + 4y = 0 \\ 2x^2 + y = 0 \end{array}\right.$

If we add them, we have $5y = 0 \rightarrow y = 0$.

Later, we substitute $y = 0$ in the first equation, for example, and we obtain $x = 0$.

So, solution is $\boxed{x = y = 0}$.

---

63. Solve the following systems:

a) $\left.\begin{array}{l} x + y = 1 \\ x^2 - 2x + 3y = -1 \end{array}\right\}$

b) $\left.\begin{array}{l} xy = 12 \\ 3x - 2y = 1 \end{array}\right\}$

c) $\left.\begin{array}{l} x^2 + y^2 = 25 \\ x + y = 2 \end{array}\right\}$

d) $\left.\begin{array}{l} x + y = 14 \\ 24 + 24y = 7xy \end{array}\right\}$

e) $\left.\begin{array}{l} x^2 - y = 0 \\ x - y = 0 \end{array}\right\}$

f) $\left.\begin{array}{l} x - y = 11 \\ y^2 = x - 5 \end{array}\right\}$

g) $\left.\begin{array}{l} x^2 + y^2 = 5 \\ x^2 - y^2 = 3 \end{array}\right\}$

h) $\left.\begin{array}{l} x + y = 2 \\ x^2 + 2y = 7 \end{array}\right\}$

i) $\left.\begin{array}{l} 2x + y = 3 \\ x^2 - y^2 = 0 \end{array}\right\}$

j) $\left.\begin{array}{l} 2x^2 + y^2 = 18 \\ 3x^2 - y^2 = -13 \end{array}\right\}$

k) $\left.\begin{array}{l} 3x + y = 1 \\ xy = -2 \end{array}\right\}$

l) $\left.\begin{array}{l} 3x^2 - 5y^2 = 30 \\ x^2 - 2y^2 = 7 \end{array}\right\}$

64. Solve the following systems:

a) $\left.\begin{array}{l} x - y = 2 \\ x^2 - y^2 = 4 \end{array}\right\}$

b) $\left.\begin{array}{l} x + 2y = 6 \\ xy = -8 \end{array}\right\}$

c) $\left.\begin{array}{l} x + y = 4 \\ x^2 - xy = 6 \end{array}\right\}$

65. Solve the following systems:

a) $\left.\begin{array}{l} x - y + 3 = 0 \\ x^2 + y^2 = 5 \end{array}\right\}$

b) $\left.\begin{array}{l} x + y = 1 \\ xy + 2y = 2 \end{array}\right\}$

c) $\left.\begin{array}{l} 2x + y = 3 \\ xy - y^2 = 0 \end{array}\right\}$

d) $\left.\begin{array}{l} 3x + 2y = 0 \\ x(x - y) = 2y^2 - 8 \end{array}\right\}$

66. Solve the following systems:

a) $\left.\begin{array}{l} x^2 + y^2 = 41 \\ x^2 - y^2 = 9 \end{array}\right\}$

b) $\left.\begin{array}{l} 3x^2 + 2y^2 = 35 \\ x^2 - 2y^2 = 1 \end{array}\right\}$

c) $\left.\begin{array}{l} x^2 + y^2 + x + y = 32 \\ x^2 - y^2 + x - y = 28 \end{array}\right\}$

d) $\left.\begin{array}{l} x^2 + 2y^2 + x + 1 = 0 \\ x^2 - 2y^2 + 3x + 1 = 0 \end{array}\right\}$

Exercises

7. Inequalities

Inequalities are expressions quite similar to equations, but instead of expressing the equality of both members by the symbol "=", their inequality is expressed by $<$, $>$, $\leq$ or $\geq$.

		symbol	Graphically
$\geq$ bigger or equal than $\leq$ lower or equal than	Values are included	$\bullet$	[ ]
$>$ bigger than $<$ lower than	Values are NOT included	o	( )

Inequalities are solved by the same steps than equations. We must only be careful in one aspect:

When **multiplying** or **dividing** both members of an inequality by a **negative** number, the symbol of the inequality must be **changed**.

For example: $2x + 5 < 5x + 20 \;\rightarrow\; 2x - 5x < 20 - 5 \;\rightarrow\; (-3)x < 15 \;\rightarrow\; x > \dfrac{15}{(-3)} \;\rightarrow\; \boxed{x > (-5)}$.

Now, some inequalities are solved, step by step, for you to understand the process:

Example: Solve: $2(x + 1) - 3(x - 2) < x + 6$

Solution: $2x + 2 - 3x + 6 < x + 6$

$2x - 3x - x < -2 - 6 + 6$

$-2x < -2 \;\rightarrow\; \boxed{x > 1}$ or $(1, +\infty)$ or

Example: Solve: $\dfrac{3x+1}{7} - \dfrac{2-4x}{3} \geq \dfrac{-5x-4}{14} + \dfrac{7x}{6}$

Solution: LCM(7, 3, 14, 6) = 42. So, we multiply all the terms by 42:

$6(3x + 1) - 14(2 - 4x) \geq 3(-5x - 4) + 49x$

$18x + 6 - 28 + 56x \geq -15x - 12 + 49x$

$18x + 56x + 15x - 49x \geq -12 - 6 + 28$

$40x \geq 10 \rightarrow 4x \geq 1 \rightarrow \boxed{x \geq 1/4}$ or $[1/4, +\infty)$

67. Solve the following linear inequalities:

a) $x + 2x + 3x < 5x + 1$

b) $5x + 10 > 12x - 4$

c) $4x + 2 - 2x < 8x$

d) $2x + 4 > x + 6$

e) $-x + 1 > 2x + 4$

f) $5x + 10 < 12x - 4$

68. Solve the following linear inequalities:

a) $x + 2x + 3x > 5x + 1$

b) $5x + 10 < 12x - 4$

c) $4x + 2 - 2x > 8x$

d) $2x + 4 > x + 6$

e) $-x + 1 < 2x + 4$

f) $x + 51 > 15x + 9$

69. Find out the numbers whose triple minus 20 units is lower than its double plus 40 units.

70. Solve the following linear inequalities:

a) $x + 2x + 3x < 5(1 - x) + 6$

b) $(x - 1) + 2(2x + 3) < 4$

c) $6(x - 2) - 7(x - 4) > 6 - 3x$

d) a) $x + 2x + 3x > 5(1 - x) + 6$

e) $-1(x - 1) + 2(2x + 3) > 4$

f) $6(x - 2) - 7(x - 4) < 6 - 3x$

71. The price for mobile calls at the firm A is fix 20 euros/month plus 7 cents/min, and in the firm B it is fix 11 euros/month plus 12 cents/min. For more than how minutes of conversation is more convenient form A?

72. Solve the following linear inequalities:

a) $2(x - 3) > 1 - 3(x - 1)$

b) $2(x + 1) + 4 < -2(x + 3)$

c) $(x - 20) / 8 < (1 - 2x) / 10$

d) $\dfrac{x - 1}{4} < \dfrac{x + 3}{3} - \dfrac{x - 5}{2}$

73. A man is 25 years older than his son. Find out the period of their lives in which father's age exceeds in more than 5 years the double of the age of his son.

Quadratic inequalities

Quadratic inequalities are solved following the same steps than quadratic equations. The only difference is, at the end, when looking for the intervals that are solution of the inequality. We are seeing it with some examples:

Example: Solve: $x^2 + 3x - 4 > 0$

Solution: We are solving it as though it were a quadratic equation: $x^2 + 3x - 4 = 0$

$$x = \frac{(-3) \pm \sqrt{9 + 16}}{2} = \frac{-3 \pm 5}{2} \quad \rightarrow \quad \begin{cases} x_1 = 1 \\ x_2 = (-4) \end{cases}$$

At this moment, we have three zones in the Real Line:

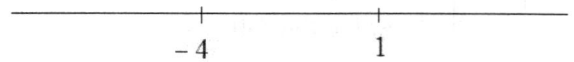

Now, we must decide which of these zones fulfil the condition $x^2 + 3x - 4 > 0$

We one value included in each zone:

$P(-6) = 36 - 18 - 4 = 14 > 0 \rightarrow$ This zone fulfils the inequality.

$P(0) = -4 < 0 \rightarrow$ This zone does not fulfil the inequality.

$P(10) = 100 + 30 - 4 = 126 > 0 \rightarrow$ This zone fulfils the inequality.

IMPORTANT: See that (-4) and 1 do NOT fulfil the inequality, as $P(-4) = P(1) = 0$

So, solution of this inequality is: $(-\infty; -4) \cup (1; +\infty)$ or

Example: Solve: $-x^2 + 4x - 7 < 0$

<u>Solution:</u> We are solving it as though it were a quadratic equation: $x^2 - 4x + 7 = 0$

$$x = \frac{4 \pm \sqrt{16 - 28}}{2} = \frac{4 \pm \sqrt{-12}}{2} \rightarrow$$

As you know, this equation has NOT any solution. So, Real Line is not divided into several zones, we must

consider it as an only zone $(-\infty; +\infty)$. So, we will consider only one value: For example, $P(0) = -7 < 0$.

So, the whole Real Line is solution of this inequality. Sol.: $x = R$.

---

74. Solve the following quadratic inequalities:

a) $4x^2 - 2x < 2$　　　b) $5x^2 - 6x + 1 \geq 0$　　　c) $3x^2 < -4x + 4$　　　d) $(4x - 8)(x + 3) < 0$

e) $(x + 1)(2x + 1) \geq 0$　　　f) $-x^2 - x + 3 < 0$　　　g) $\dfrac{x^2 + x}{3} - 1 > -\dfrac{1 - 2x^2}{6}$

75. Solve the following quadratic inequalities:

a) $\dfrac{2x^2}{3} - x < \dfrac{8x}{3}(1 + x) + 1$　　　b) $\dfrac{x - 4}{3} < \dfrac{x^2}{x + 42}$　　　c) $\dfrac{x + 2}{3} < \dfrac{x^2}{3x + 4}$

d) $x^2 + 2x + 3 \leq -1$　　　e) $(x + 5)(x - 4) \geq 0$　　　f) $x^2 - 9 < 0$　　　g) $-(x + 2)(x - 6) \leq 0$

Exercises

Systems of inequalities with one unknown

Systems of inequalities are formed by two inequalities and their solution is the zone that fulfils both of them at the same time.

Example: Solve the following system of inequalities:

$$\begin{cases} 10(x+1) + x \le 6(2x+1) \\ 4(x-10) < -6(2-x) - 6x \end{cases}$$

<u>Solution:</u> We will solve both inequalities separately:

$10(x+1) + x \le 6(2x+1)$

$10x + 10 + x \le 12x + 6$

$10x - 12x \le -10 + 6$

$-2x \le -4 \;\rightarrow\; 2x \ge 4 \;\rightarrow\; x \ge 2$

$4(x-10) < -6(2-x) - 6x$

$4x - 40 < -12 + 6x - 6x$

$4x < 40 - 12 \;\rightarrow\; 4x < 28 \;\rightarrow\; x < 7$

In the following figure, you can see there is a zone that has been marked twice.

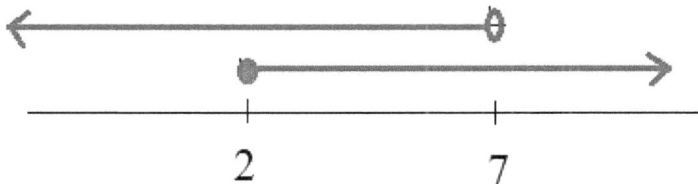

And this is the solution of this system: $[2, 7)$.

---

76. Solve the following systems of inequalities:

a) $\begin{cases} 4x - 3 < 1 \\ x + 6 > 2 \end{cases}$
b) $\begin{cases} 2x - 3 > 0 \\ 5x + 1 < 0 \end{cases}$
c) $\begin{cases} 3x - 4 < 4x + 1 \\ -2x + 3 < 4x - 5 \end{cases}$

d) $\begin{cases} 3x + 1 > x + 9 \\ x + 5 < 2 - 3x \end{cases}$
e) $\begin{cases} 2x - 5 < 0 \\ x - 4 > -5 \end{cases}$

77. Solve the following systems of inequalities:

a) $\begin{cases} 2x + 2 < 6 \\ 3x - 1 \ge -7 \end{cases}$
b) $\begin{cases} x - 2 \ge 0 \\ 2x \le 10 \end{cases}$
c) $\begin{cases} 5 - x < -12 \\ 16 - 2x < 3x - 3 \end{cases}$
d) $\begin{cases} 3x - 2 > -7 \\ 5 - x < 1 \end{cases}$

Exercises

Review exercises

Linear equations

1. Solve the following linear equations:

a) $(x+1)^2 + (x-2)^2 = (x+2)^2 + (x-1)^2$

b) $4(x-3)(x+3) - (2x+1)^2 = 3$

c) $(x-3)^2 + 1 = (x+2)^2 - 4x - 3(x-1)$

d) $5(x-3)^2 + x^2 - 46 = -(2x+1)(1-3x)$

e) $(4x-3)(7x+2) - (3-4x)^2 = 3x(4x-5) - 2$

2. Solve the following linear equations:

a) $2(2-x) - 4(2x-1) = 4(x-1) - 3(2x-3)$

b) $\dfrac{5x-2}{6} + \dfrac{2x+1}{9} = x$

c) $4(2-3x) - 2(1-3x) = 2(3x-2) - 5(x-4)$

d) $x - 1 - \dfrac{x-2}{2} + \dfrac{x+3}{3} = 0$

e) $3\left(\dfrac{x}{2} - 2\right) + 4\left(\dfrac{3x}{2} - 1\right) = 5(3x-8)$

f) $\dfrac{3}{2}\left(x - \dfrac{1}{3}\right) - \dfrac{5}{3}\left(\dfrac{2x}{3} + 2\right) = \dfrac{x-3}{6}$

g) $\dfrac{2x-3}{3} - \dfrac{4x-7}{10} = \dfrac{3x-8}{15} + \dfrac{x-4}{6}$

h) $\dfrac{1}{2}\left(\dfrac{2x-9}{9} - \dfrac{x+4}{4}\right) + \dfrac{5x+6}{4} = \dfrac{7x-2}{4}$

i) $\dfrac{3-x}{5} - \dfrac{2(x+4)}{3} - (x-1) = \dfrac{x}{10}$

j) $x + 3 - \dfrac{3(2x-5)}{4} = 3x - \dfrac{5(2-3x)}{7}$

Quadratic equations

3. Solve the following quadratic equations:

a) $6 - 9x^2 - 15x = 0$

b) $x^2 + 11x = 0$

c) $3x^2 - 6 = 21$

d) $3 - 4x^2 = 3(1 + 4x^2)$

e) $4x(x+2) - 5 = 12 - (x-4)^2$

f) $x(x-2) + 3x = 2x^2$

g) $4x^2 - 9 = 0$

h) $x^2 - \dfrac{x}{2} = \dfrac{1}{3} - \dfrac{2x}{3}$

i) $2x^2 + \dfrac{6}{5} = x\left(x + \dfrac{31}{5}\right)$

j) $\dfrac{(x+1) + (x+2)\cdot(x+3)}{4} - (x+1) = \dfrac{11x+2}{6}$

k) $\dfrac{x^2 + 2x + 1}{2} - \dfrac{x+1}{4} = 9$

l) $(x+2)^2 = 2x(x+2) + 4(x+1)$

m) $\dfrac{(x-1)^2 - 3x + 1}{15} + \dfrac{x+1}{5} = 0$

n) $x + 6 + x(x-2)^2 = x(x-1)^2$

ñ) $\dfrac{x}{3}(x-1) - \dfrac{x}{4}(x+1) + \dfrac{3x+4}{12} = 0$

o) $(x+1)^2 - (x-2)^2 = (x+3)^2 + x^2 - 20$

p) $\dfrac{x^2 - 2x + 5}{2} - \dfrac{x^2 + 3x}{4} = \dfrac{x^2 - 4x + 15}{6}$

Biquadratic equations

4. Solve the following biquadratic equations:

a) $x^4 - 10x^2 + 9 = 0$

b) $x^4 - \dfrac{13}{36}x^2 + \dfrac{1}{36} = 0$

c) $(2x^2 + 1)(x^2 - 3) = (x^2 + 1)(x^2 - 1) - 8$

d) $4x^4 - 17x^2 + 4 = 0$

e) $\dfrac{1}{4}(3x^2 - 1)(x^2 + 3) - \dfrac{1}{3}(2x^2 + 1)(x^2 - 3) = 4x^2$

f) $x^2(x^2 - 17) + 16 = 0$

5. Solve the following equations:

a) $x^3 - x^2 - 4x + 4 = 0$

b) $x^4 + 2x^3 = x^2 + 2x$

c) $x(x^2 + 2x - 1) = 2$

d) $x^4 + 3x^3 - x^2 - 3x = 0$

e) $(x + 1)^3 + 3x(x^2 - 5) = 7x^2 + 4x - 15$

f) $125 - x^3 = 0$

g) $(2x^2 + x - 1)(4x - 1) = 8x - 2$

h) $x^4 - 7x^2 + 6x = 0$

i) $x(x^2 - 3x + 4) = 2(x + 3)$

j) $(x^2 - 1)^2 - x(x^2 - x + 1) = 3$

k) $(3x^2 - 1)(2x - 1) = 7x - 2x^2 - 3$

l) $x^4 = 4x^2$

6. Solve the following equations:

a) $\dfrac{x}{7} + \dfrac{21}{x+5} = \dfrac{47}{7}$

b) $\dfrac{x}{x+1} + \dfrac{x}{x+4} = 1$

c) $\dfrac{15}{x} - \dfrac{72 - 6x}{2x^2} = 2$

d) $\dfrac{x+1}{x-1} - \dfrac{2-x}{x} = 3$

e) $\dfrac{x-3}{x^2 - x} - \dfrac{x+3}{x^2 + x} = \dfrac{2 - 3x}{x^2 - 1}$

f) $\dfrac{3x}{x^2 - 9} = 1 + \dfrac{x}{2(x-3)}$

g) $\dfrac{3x+1}{x-2} + 4 = \dfrac{x}{x+2} + \dfrac{x^2}{x^2 - 4}$

h) $\dfrac{8-x}{2} - \dfrac{2x-11}{x-3} = \dfrac{x+6}{2}$

i) $\dfrac{4}{x+2} + \dfrac{3x}{x-2} = \dfrac{3x^2 - 8}{x^2 - 4}$

j) $\dfrac{x}{x-3} - \dfrac{2x+1}{x^2 - 9} = \dfrac{1}{x+3}$

7. Solve the following equations:

a) $\dfrac{(x-3)^2}{4} - \dfrac{(2x-1)^2}{16} = \dfrac{35}{16}$

b) $\dfrac{x+3}{5} + \dfrac{(x-1)^2}{4} = \dfrac{1}{4}x^2 - \left(\dfrac{x}{2} + 2\right)$

c) $\dfrac{1}{2}\left[1 - (x+2)^2\right] = -x - \dfrac{x^2 - 1}{2}$

d) $\dfrac{1}{2}\left[2(x+1) - (x-3)^2\right] = \dfrac{1}{2}\left[3(x-1) - \dfrac{2}{3}(x+1)^2\right]$

e) $\dfrac{1 - 2x}{9} = 1 - \dfrac{x+4}{6}$

f) $\dfrac{3x+2}{5} - \dfrac{4x-1}{10} + \dfrac{5x-2}{8} = \dfrac{x+1}{4}$

g) $(3x + 2)^2 + 3(1 - 3x)x = 2(x - 11)$

h) $(2x - 3)^2 + (x - 2)^2 = 3(x + 1) + 5x(x - 1)$

Problems with equations and systems

8. When renting a car in a firm, you have to pay per day and kilometre. One customer paid 160 € for 3 days and 400 km, and another one paid 175 € for 5 days and 300 km. Find how much you have to pay per day and per km.

77

9. Someone bought two pictures as an investment for 2650 €. Two years later, he sold them for 3124 €, being the first one 20% more expensive than originally, and the second one, 15% more expensive. What were their original prices?

10. A chemist has two types of commercial alcohol, containing one of them 80% of alcohol, and the other one, 95%. How many litres has he to mix if he needs 5 litres containing 86% of alcohol?

11. Someone bought two motorbikes for 3000 € and he sold them for 3330 €. What were their original prices if the first one was sold 25% more expensive and the second one was sold 10% cheaper?

12. A mixture of 5 kg of green painting and 3 kg of white one costs 69 €. Find out the Price of one kilogram of them, knowing that a mixture containing one kg of each one would cost 15 €.

13. Find out the sides of a rectangle knowing its perimeter, 34 m, and its area, 60 m^2.

14. The perimeter of an isosceles triangle is 32 cm and the height on its different side is 8 cm. Find out the sides of the triangle.

15. The total area of a cylinder is 112π cm^2 and its radius and height together measure 14 cm. Find out its volume.

16. If we increase in 5 cm the side of a square, its area is multiplied by 4. What was the original side of the square?

17. The perimeter of a right triangle is 36 m and one leg is 3 cm shorter than the other one. Find out the sides of this triangle.

18. One person takes 3 hours more than another one to do the same work. If they do the work together, it takes them 2 hours. How long does it take to each one, separately?
☞ *If someone takes x hours to do a work, in one hour, he will have done 1/x of the work.*

19. A tap takes a double time to fill a container than another one. If we open both taps together, the container gets full in 3 minutes. How long does it take to each tap to fill the container?

20. A group of friends rent a van for a trip and it costs 490 €. The previous day to leaving, two more friends join them and they have 28 € given back. How many people did take part in the trip and how much did the pay?

21. The total price of the computers of a shop is 60 000 €. But two of them get broken and he must increase the Price of the other ones in 50 € in order to get the same benefit. How many computers did he have and what was their final price?

22. A family went for a 300 km trip. On the return, their speed was 10 km/h higher and it has taken them one hour less. Find out the speeds and the times for both parts of the trip.

23. We have a rectangle shaped field. If its base decreased in 80 m and its height increased in 40 m, it would become a square. If its base decreased in 60 m and its height increased in 20 m, its area would decrease in 400 m^2. What are the measures of the field?

24. Find out the dimensions of a rectangle whose diagonal measures 13 m, and its area is 60 m^2.

25. The side of a rhombus is 5 cm, and its area is 24 cm^2. Find out the length of its diagonals.

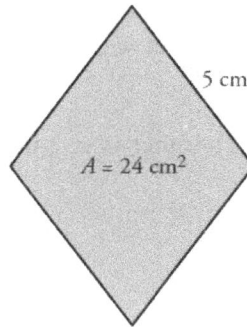

26. The addition of the two digits of a number is 8. If we add 18 units to this number, we will obtain the same number but with its digits in the inverse order. What is this number?

27. The two digits of a number are consecutive numbers. If we divide this number by the one resulting after changing the order of its digits, the quotient is 1.2. What is the number?

Inequalities

28. Solve the following linear inequalities:

a) $2 (x - 1) + 5 < 3 x - 1$

b) $3 x - (x - 1) - 7 \geq 5 (x - 1)$

c) $2 (x - 1) - 3 (x + 1) \leq 2 (2 - x)$

d) $3 - \dfrac{x + 1}{2} < \dfrac{x + 1}{5}$

e) $\dfrac{x - 1}{3} - \dfrac{x + 2}{6} \leq \dfrac{2 x + 1}{4} - \dfrac{x + 3}{9}$

f) $2 \cdot \dfrac{x + 1}{5} - \dfrac{3 \cdot (2 x - 1)}{2} \geq \dfrac{x + 1}{10} - 3 x$

29. Solve the following linear inequalities:

a) $2(x - 3) < 1 - 3(x - 1)$

b) $10(20 - x) > 8(2x - 1)$

c) $2(1 - x) - 4 < 2(x + 3)$

d) $-4x + \dfrac{3 - 2x}{4} > \dfrac{1 - 3x}{3} - \dfrac{37}{12}$

30. Solve the following quadratic inequalities:

a) $x^2 - 2x + 1 \leq 3 (x - 1)$

b) $x^2 - 6x + 9 < 0$

c) $x^2 - 3x - 10 > 0$

d) $(x-2)(x+3) \geq 0$

e) $x(x + 3) > 2 - x^2$

f) $(x + 1)(x - 1) \geq 0$

g) $\left(\dfrac{x}{2} + 3 \right)(-x + 1) > 0$

h) $x^2 - 2x - 3 \geq 0$

i) $4x^2 + 4x + 3 < 0$

j) $(x - 1)(x - 6) \leq 0$

31. In a 40 questions exam, you win two marks per correct answer, but your qualification decreases 0.5 marks per mistake. You must answer all the questions. What is the minimum number of correct answers you need if you want to have a qualification of, at least, 40 marks?

32. The product of a whole number by another one, 2 units bigger, is less than 8. What can be that number?

33. If we subtract the squared of a number minus 3 times itself, we obtain more than 4. What can you say about this number?

34. A group of friends have gathered 50 € to go to a party. If the ticket is 6 €, the have more money than necessary, but if it is 7 €, they do not have enough money. How many friends are they?

Systems of inequalities

35. Solve the following systems of inequalities:

a) $\begin{cases} 6x+5 \le 5x-2 \\ 3x-\dfrac{1}{2} > -5 \end{cases}$
b) $\begin{cases} x+1 < 2 \\ x-1 > -2 \end{cases}$
c) $\begin{cases} 2x+1 > x-\dfrac{3}{2} \\ 2x-1 < 1-3x \end{cases}$
d) $\begin{cases} x-\dfrac{1}{3} < \dfrac{3}{2}x-1 \\ 4x-5 < 2-5x \end{cases}$

36. Solve the following systems of inequalities:

a) $\begin{cases} 6-x \le 4x-5 \\ 1-2x \ge -3 \end{cases}$
b) $\begin{cases} 2x-6 < 0 \\ x-4 > -5 \end{cases}$
a) $\begin{cases} 2x-15 \le x-5 \\ -x+12 \ge 6 \end{cases}$
b) $\begin{cases} 2x-10 > -x+2 \\ 10-4x > -3x \end{cases}$

37. Solve the following system of inequalities:

$$\begin{cases} 8x-7 > \dfrac{15-9x}{2} \\ 4x-5 > 5x-\dfrac{8}{3} \end{cases}$$

38. Solve:

a) $x^3 - 2x^2 - 2x - 3 = 0$
b) $2x^3 - 7x^2 - 19x + 60 = 0$
c) $x^3 - x - 6 = 0$
d) $4x^4 + 4x^3 - 3x^2 - 4x - 1 = 0$

$$f(x)=2^x$$

$$g(x)=\log_2 x$$

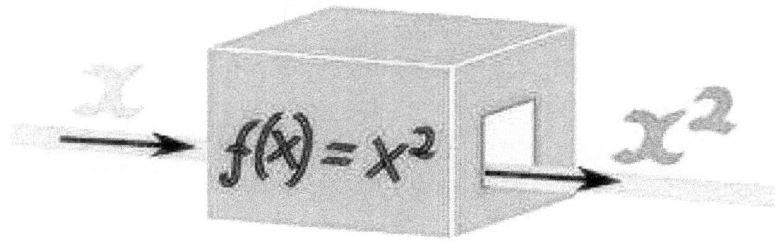

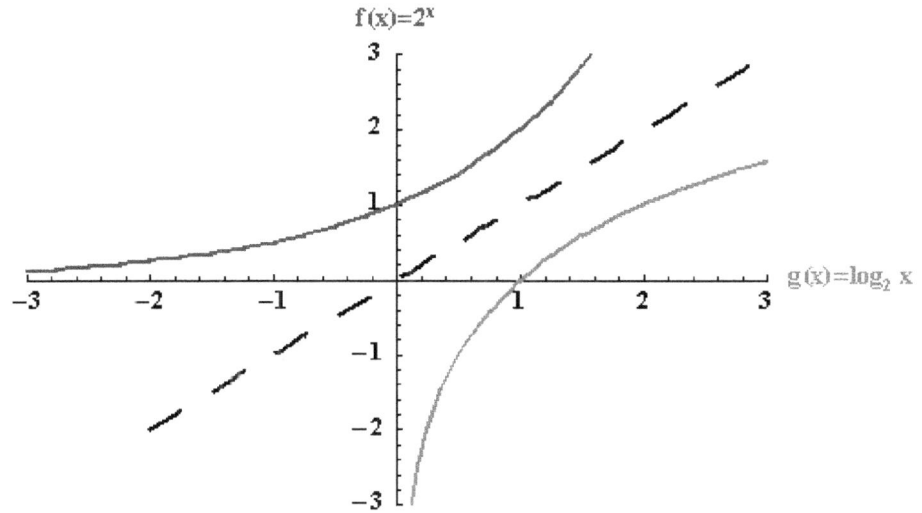

Unit 4.- Functions and graphs

1. First concepts

As functions are usually graphically represented, two axes are needed, one of them horizontal and the other one, vertical. They will be the reference axes to locate points in the plane $\mathbf{R}^2$. They are named *X-axis* or *abscise axis*, and *Y-axis* or *ordinated axis*.

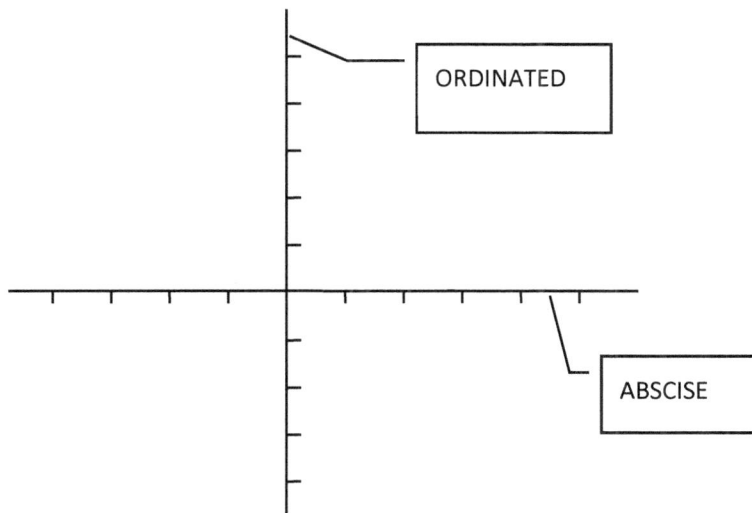

ORDINATED

ABSCISE

Functions express the relationship between two magnitudes or variables, so that each value of the independent variable *x*, corresponds *an only value* of the dependent variable *y*.

In order to indicate that a magnitude *(y)* depends or is a function of another one *(x)*, it is used the notation *y = f(x),* which is read *"y is a function of x"*.

In order to visualize the behavior of the function, we use its graphical representation: we represent both variables on two coordinated axes, including their scales:

· x variable is represented on the horizontal axis or **abscise axis.**
· y variable is represented on the vertical axis or **ordinated axis.**

2. How are functions given to us?

a) Value tables

In a laboratory class, a student has measured temperature of a liquid while it was being heated. He wrote results in the following table:

x = time (min)	0	1	2	3	...
y = temperature (°C)	20	24	28	32	...

Temperature of the liquid (*dependent variable*) depends on the time (*independent variable*).

b) Graphs

Following graph represents variation of temperature of an ill person in a hospital during a whole day. From it, it is easy o determine the relationship between the time and the temperature of that person. For example, his lowest temperature has been reached at 2 p.m. (14:00 h) and highest one at 8 p.m. (20:00 h).

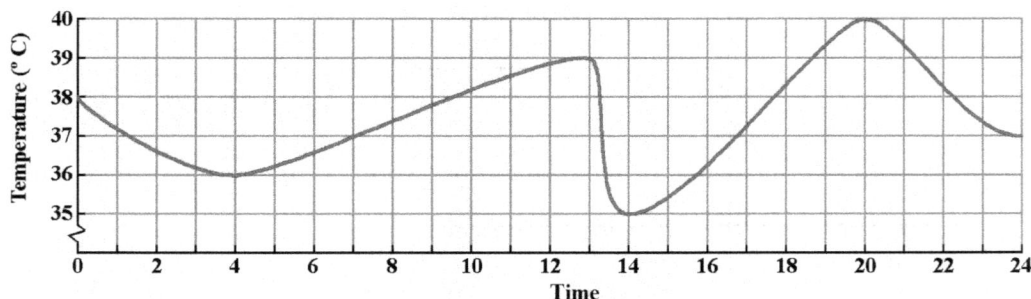

Graph gives us an intuitive view about the relationship between two magnitudes and the behavior of one magnitude, depending on the other one. Again, temperature (*dependent variable*) depends on the time in the day (*independent variable*).

c) Algebraic expressions

Many of the relationships between magnitudes in Nature are given by algebraic expressions.

- Volume of a sphere depends on its side. The formula that relates both magnitudes is $V = \dfrac{4}{3}\pi a^3$.

- If price of 1 kg of oranges is 0.35 €, the cost *(C)* of a quantity of oranges depends on the number *(k)* of kilograms. The formula that relates both magnitudes is $C = 0.35k$.

1. Decide which of the following curves are functions and explain why the others are not.

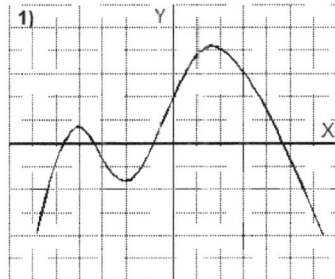

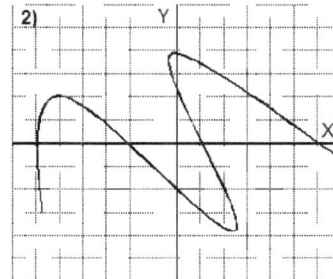

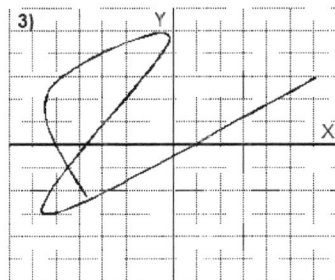

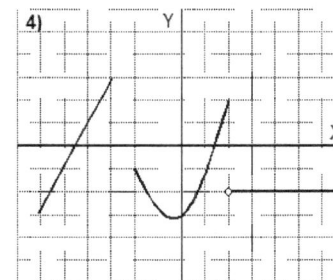

2. The following curve represents one object´s temperature during a period of time:

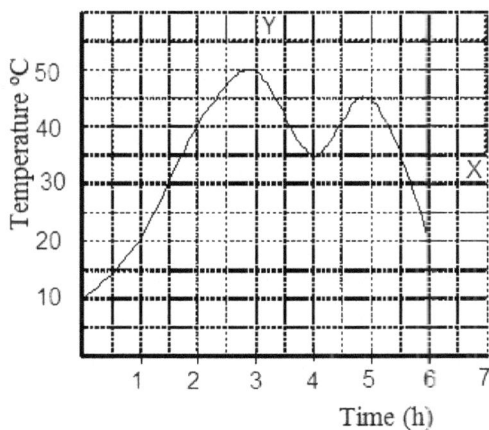

a) Which are the variables?

b) Between what values are variables located?

83

3. Find out the image of these abscises:

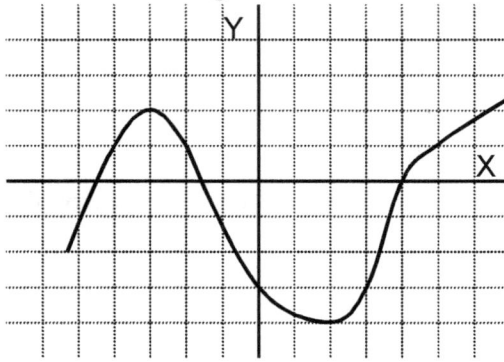

a) f(2) =

b) f(3) =

c) f(0) =

d) f(4) =

e) f(-2) =

f) f(-4) =

a) f(1) =

b) f(0) =

c) f(-1) =

d) f(2) =

e) f(-3) =

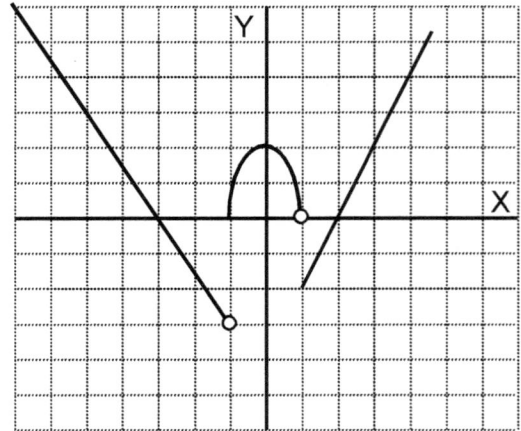

4. Write the function that gives the product of two numbers, knowing their addition is 18.

5. Express the area of an equilateral triangle as a function of its side (x).

6. We have a rectangular paper whose dimensions are 5 dm x 3 dm, and we have cut a little square in each corner, in order to have an open box. If the side of the little square is x, find out the voume of the box as a function of x.

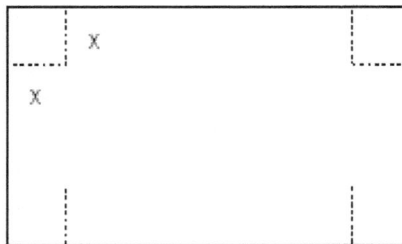

84

7. From this values chart, represent these points and join them:

x	f(x)
-1	2
0	1
1	0
2	-1
3	-2

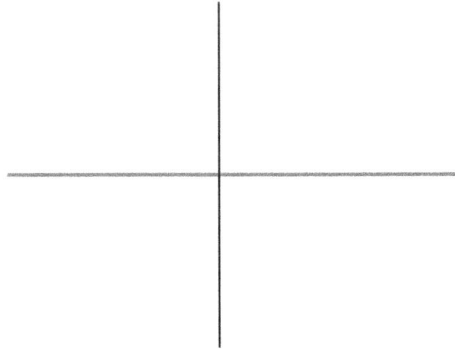

3. Intersection points with the axes

When studying functions, it is interesting knowing if they intersect axes and where they do it.

- Intersection points with ordinated axis
Intersection point of a function $y = f(x)$ with the Y-axis has abscise = 0. So, its coordinates are **$(0, f(0))$**.

- Intersection points with abscise axis
Intersection point of a function $y = f(x)$ with the X-axis has ordinated 0. So, they are the points whose x coordinate are solution of equation $f(x) = 0$.

Example: Find out the intersection points of function $y = \dfrac{3}{5}x - 3$ with the coordinated axes.

<u>Solution:</u>   For $x = 0 \rightarrow \quad y = -3 \quad \rightarrow \quad Q(0, -3)$

For $y = 0 \rightarrow \quad \dfrac{3}{5}x - 3 = 0 \quad \rightarrow \quad x = 5 \quad \rightarrow \quad P(5, 0)$

Look at the plot of the function:

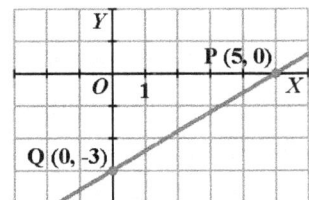

8. Write the coordinates of five points located on the abscise axis. What do they have in common?

9. Write the coordinates of five points located on the ordinated axis. What do they have in common?

10. Find out the intersections with the axes for these functions:

a) $y = 2x - 1$ b) $y = 3x + 2$ c) $y = -3x + 2$ d) $y = 2$ e) $y = x^2 - 2x - 3$

f) $y = x^2 - 2x + 1$ g) $y = 4 - x^2$ h) $2y = x - 4$ i) $3y - 6 = 2x$

Continuity of a function is a very intuitive idea. A function will be said to be a continuous one if it can be drawn by an only stroke. In the same way, one function will be said to be a discontinuous one if it has some discontinuity.

Definition of continuity

A function is said to be a **continuous** one if it does not have any kind of disccontinuity. A function can be a **continuous in an interval** if it has discontinuities only outside that interval.

Functions are continuous into their domains.

Types of discontinuities

There are several reasons by which a function can be discontinuous at one point:

• **Asymptotic discontinuity:** It has infinite vertical asymptotes at that point. This means y values of the function increase or drop infinitely when x coordinate gets closer and closer to that point.

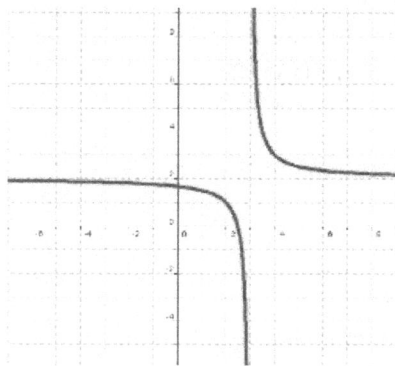

• **Jump discontinuity:** When moving along the curve of the function, we come across a jump.

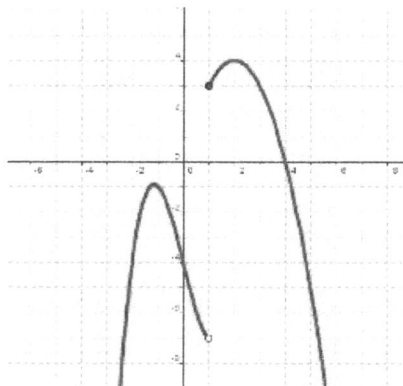

• **Point discontinuity**: At one point, the function is not defined or it is defined at a point outside the curve.

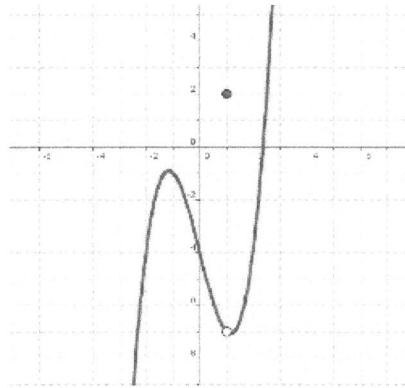

Notice a point discontinuity can be easily "repaired" by only adding the correct point in its definition:

For example, observe the curve of the function defined as $f(x) = \begin{cases} x \ if \ x < 1 \\ 3 \ if \ x = 1 \\ x \ if \ x > 1 \end{cases}$

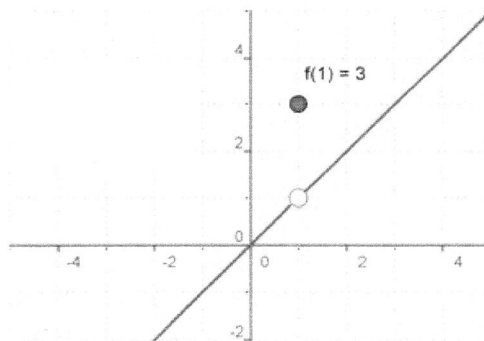

f(1) = 3

This function has a point discontinuity at x = 1, but it can be "repaired" if we change the 3 by a 1 in its

definition: $f(x) = \begin{cases} x \ if \ x < 1 \\ 1 \ if \ x = 1 \\ x \ if \ x > 1 \end{cases}$ Now, it is a continuous function.

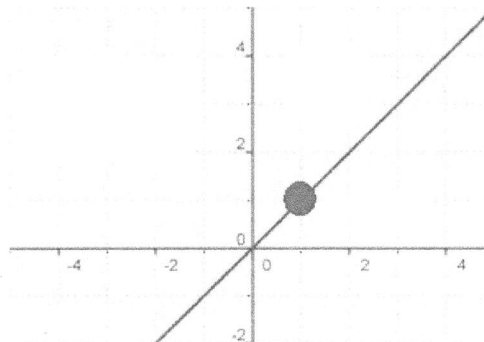

That is why point discontinuities are named *avoidable discontinuities*.

$$\text{Types of discontinuities} \begin{cases} \text{Unavoidable discontinuities} \begin{cases} \text{First kind} \to \text{Asymptotic discontinuities} \\ \text{Second kind} \to \text{Jump discontinuities} \end{cases} \\ \text{Avoidable discontinuities} \to \text{Point discontinuities} \end{cases}$$

5. Variations of a function. Maximums and minimums. Average rate of change

5.1. Increasing and decreasing

> · A function $y = f(x)$ is **increasing** when if increasing independent variable, dependent variable increases as well.   **if $x_1 < x_2 \to f(x_1) < f(x_2)$**
> · A function $y = f(x)$ is **decreasing** when if increasing independent variable, dependent variable decreases.   **if $x_1 < x_2 \to f(x_1) > f(x_2)$**
> · A function $y = f(x)$ is **constant** when if increasing independent variable, dependent variable does not change.   **if $x_1 < x_2 \to f(x_1) = f(x_2)$**

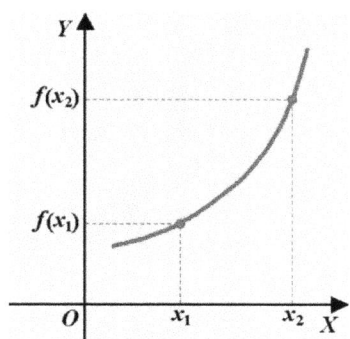

Increasing	**Decreasing**	**Constant**
$x_1 < x_2 \to f(x_1) < f(x_2)$	if $x_1 < x_2 \to f(x_1) > f(x_2)$	if $x_1 < x_2 \to f(x_1) = f(x_2)$

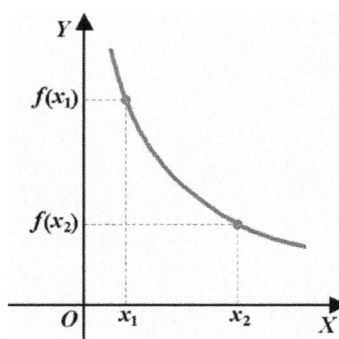

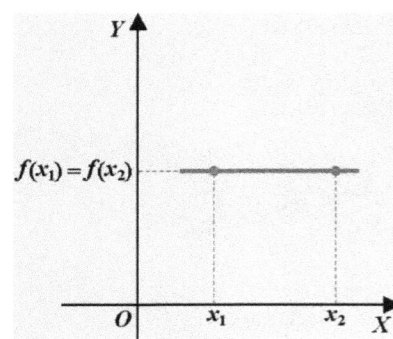

5.2. Extremes. Relative maximums and minimums. Absolute maximum and minimum.

A function $y = f(x)$ has a **relative maximum** in a point $x = x_0$ if, in values near it, to the left of x_0, $(x < x_0)$, function is increasing, and to the right of x_0, $(x > x_0)$, function is decreasing. This means that in points near it, function takes values always lower than it.

A function $y = f(x)$ has a **relative minimum** in a point $x = x_0$ if, in values near it, to the left of x_0, $(x < x_0)$, function is decreasing, and to the right of x_0, $(x > x_0)$, function is increasing. This means that in points near it, function takes values always higher than it.

A function $y = f(x)$ has an **absolute maximum** in a point $x = x_0$ if that is the highest point of its curve.

A function $y = f(x)$ has an **absolute minimum** in a point $x = x_0$ if that is the lowest point of its curve.

You are going to see it more clearly in the following figures.

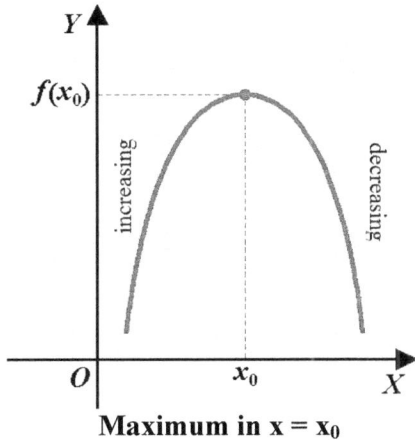

Maximum in x = x₀

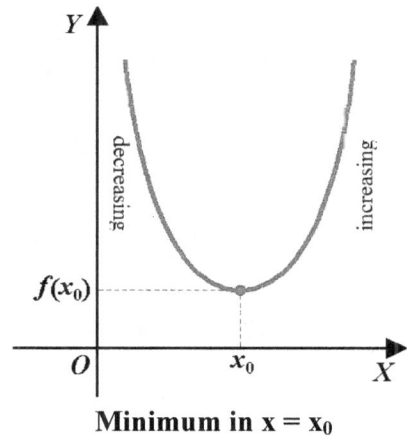

Minimum in x = x₀

Anyway, a function can have several minimums and maximums. In order to distinguish them, we define the following concepts.

Example: Next graph shows the profile of a day at the *"Tour de France"*. Study the increasing and decreasing of the function and its maximums and minimums.

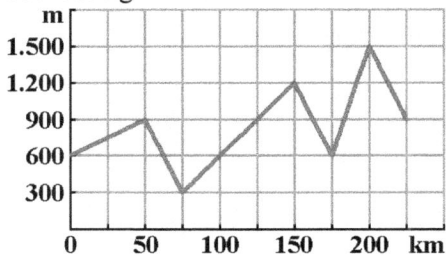

Solution:

· Function is increasing in the intervals $[0, 50)$, $(75, 150)$ and $(175, 200)$.
· Function is decreasing in the intervals $(50, 75)$, $(150, 175)$ y $(200, 225]$.

· It has an absolute maximum at x = 200 km (maximum height is 1500 m) and an absolute minimum at x = 150 km (minimum height is 1200 m).
· Relative maximums are reached at x = 50 km (h = 900 m) and x = 150 km (h = 120 m).
· Relatives minimums are reached at x=0km (h=600m), x=175km (h=600m) and x=225km (h=900m).

5.3. Average rate of change.

The average rate of change of a quantity is the change in the value of the quantity divided by the elapsed time.

For a function, the average rate of change (A.R.C) is the change in the y-value divided by the change in the x-value for two distinct points on the graph.

$$A.R.C\ [a,b] = \frac{f(b) - f(a)}{b - a}$$

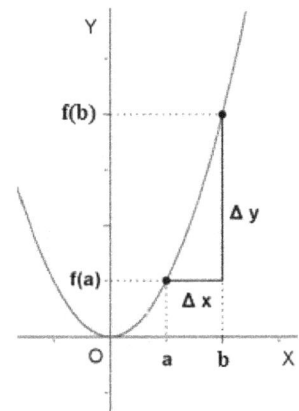

Example: Given de function $f(x) = x^2 - 3x + 5$, calculate the average rate of change between the abscise values $x_1 = (-1)$ and $x_2 = 3$.

Solution: $f(x_1) = f(-1) = (-1)^2 - 3 \cdot (-1) + 5 = 1 + 3 + 5 = 9$

$f(x_2) = f(3) = 3^2 - 3 \cdot 3 + 5 = 9 - 9 + 5 = 5$

$$A.R.C \ [(-1),3] = \frac{f(3) - f(-1)}{3 - (-1)} = \frac{9 - 5}{4} = \boxed{1}.$$

Exercises

11. Calculate the average rate of change of the function: $y = x^2 - 5x - 9$ at the following intervals: $[-2, 0], [-1, 0], [-3, -1], [0, 1]$.

12. Location of a particle is given by the function: $s = 10 + 7t - 5t^2$. Calculate the average rate of this particle at the intervals: $[2, 4], [1, 2], [1, 3], [2, 3]$.

Representation of different types of functions

6. Affine function $y = mx + n$

Exercises

13. Represent the curve of the following function: $y = 2x - 1$

x	y		Puntos:
0	$2.0 - 1 = -1$	$\longrightarrow$	$(0,-1)$
1		$\longrightarrow$	
2		$\longrightarrow$	
-1		$\longrightarrow$	
-2		$\longrightarrow$	

What kind of graph have you obtained?

Affine functions are those ones whose graph is a straight line that does not pass by the coordinate's origin. Its algebraic expression is $y = mx + n$.
In above expression:
· m is the *slope* or *gradient* of the function.
· n is the *y-intercept:* straight line intersects ordinated axis at point $(0, n)$.

14. Represent the curve of the following functions on the same coordinated axes:

$y = 5x + 2$ $y = -3x + 2$ $y = 3$

x	y

x	y

x	y

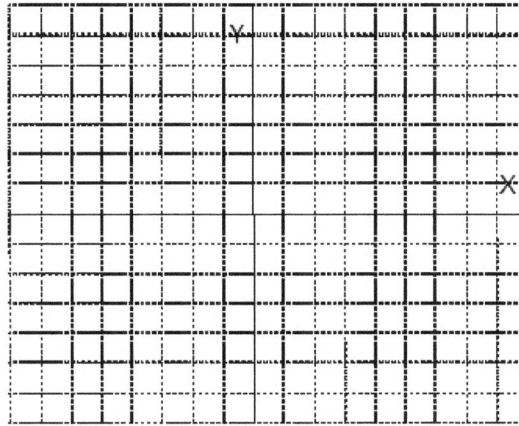

61. Affine functions having the same slope

We have drawn the straight lines corresponding to the following affine and linear functions:

[1] $y = 2x + 4$ [2] $y = 2x$ [3] $y = 2x - 6$

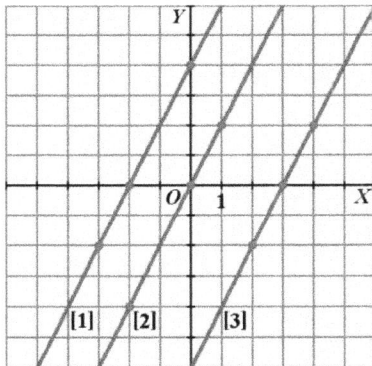

As you can see, they have the same slope, $m = 2$, and their corresponding plots are *parallel* straight lines.

> Plots of affine linear functions having **the same slope m** are **parallel straight lines**.
>
> If they have **different slopes**, they will be **secant** straight lines.

6.2. Affine functions having the same *y-intercept*

We are representing following affine functions:

[1] $y = x + 4$ [2] $y = 2x + 4$ [3] $y = -x + 4$

As you can see, all of them have 4 as their *y-intercept*, $n = 4$. This makes all straight lines pass by the point (0, 4).

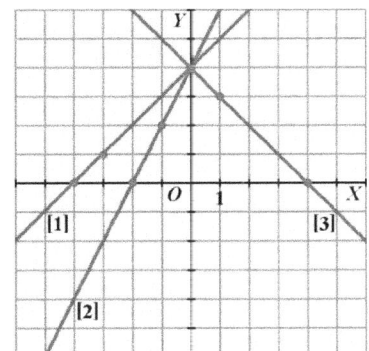

> Graphs of affine functions having **the same y-intercept n** are **secant straight lines** that intersect in the point **(0, n)**.

Exercises

15. A windows manufacturer sells his windows with a price of 3 €/m of the frame and 12 € for glass (independently on the area).

    a) How much does it cost a square window with side 2 m?

    b) Price for a window has been 60 €, how long is its side?

    c) Find the expression that gives us the price of a window as a function of the dimensions and draw it.

16. Cost of electrical energy in a house is given by the contracted power, 12 €, and the price of each consumed kilowatt·h (kw·h), that costs 0.15 €/kw·h.

    a) What is the expression that gives the cost, as a function of consume? Draw it.

    b) How much has to pay a family whose consume has been 200 kw·h?

17. Among following straight lines, indicate which are parallel and which are secant.

    a) $y = 3x + 2$ b) $y = 2x + 3$ c) $y = 3x - 3$ d) $y = 2x - 1$ e) $y = 3x - 1$

    For those being parallel, draw them on the same coordinated axes.

18. RENFE has special prices for students who want to travel through Europe in summer. A student must pay a fix tax of 30 € plus 0.02 €/km.

    a) Write the equation that relates the cost with the distance, indicating which are independent and dependent variables.

    b) Draw the plot of the function.

    c) Calculate how much must pay a student for a trip through France for 5400 kilometres.

    d) How many kilometres has a student travelled if he has paid 94 €?

19. Following table belongs to an affine function $y = mx + n$.

x	0	10	20	30	40	50
y	−3		37			97

    a) Complete the table and represent its graph.

    b) Find its algebraic expression, indicating its *slope* and *y-intercept*.

6.3. Equations of a straight line

Slope of a straight line passing by 2 points

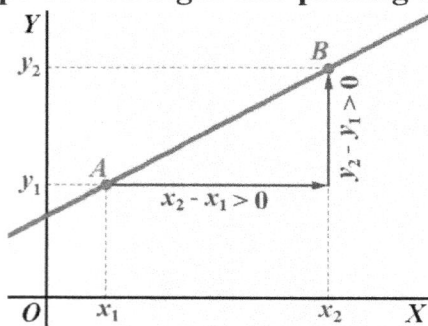

Slope, m, of a straight line passing by points $A(x_1, y_1)$ and $B(x_2, y_2)$ is calculated as

$$m = \frac{y_2 - y_1}{x_2 - x_1}$$

Example: Calculate the slope of the straight line passing by points with coordinates $A(-1, 1)$ y $B(1, 5)$.

Solution:

$$m = \frac{y_2 - y_1}{x_2 - x_1} = \frac{5 - 1}{1 - (-1)} = \frac{4}{2} = 2$$

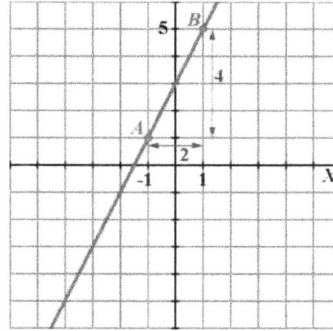

Equation of the straight line passing by 2 points

There are two methods to calculate the equation for the straight line passing by 2 points.

Example: Write the equation for the straight line passing by points with coordinates $A(-1, 1)$ y $B(1, 5)$.

Method 1:

We calculate the slope: $m = \frac{y_2 - y_1}{x_2 - x_1} = \frac{5 - 1}{1 - (-1)} = \frac{4}{2} = 2$

Affine function equation is: $y = mx + n$. So, $y = 2x + n$. We only need to know n value. We freely choose one of the points by which this function passes. For example: $B(1, 5)$.

As function passes by B, its coordinates need to verify equation of this function. We have only to

substitute: $y = 2x + n \rightarrow \begin{vmatrix} x = 1 \\ y = 5 \end{vmatrix} \rightarrow 5 = 2 \cdot 1 + n \rightarrow 5 = 2 + n \rightarrow n = 3$

So our equation is $\boxed{y = 2x + 3}$.

Method 2:

As function is passing by points A and B, their coordinates need to verify equation $y = mx + n$. If we substitute coordinates of points A and B in this equation, we will have a system of linear equations.

$A(-1,1)$: $y = mx + n \rightarrow \begin{vmatrix} x = -1 \\ y = 1 \end{vmatrix} \rightarrow 1 = m \cdot (-1) + n \rightarrow 1 = -m + n \rightarrow -m + n = 1$

$B(1,5)$: $y = mx + n \rightarrow \begin{vmatrix} x = 1 \\ y = 5 \end{vmatrix} \rightarrow 5 = m \cdot 1 + n \rightarrow 5 = m + n \rightarrow m + n = 5$

We have the following system: $\begin{cases} -m + n = 1 \\ m + n = 5 \end{cases}$ Its solution is (m = 2, n = 3).

So our equation is $\boxed{y = 2x + 3}$.

Equation of the straight line passing by points, knowing its slope or y-intercept

We are going to see it with some examples:

Example: Find the equation of the straight line being parallel to the one with equation $y = 4x - 2$, knowing it passes by the point with coordinates $(1, 9)$.

<u>Solution:</u>

The straight line we are looking for is parallel to $y = 4x - 2$. So, its slope is $m = 4$, and its equation will be $y = 4x + n$.

As it passes by point $(1, 9)$, this coordinates must verify $9 = 4 \cdot 1 + n$, so $n = 9 - 4 = 5$.

Our straight line is $\boxed{y = 4x + 5}$.

20. Work out the equation of the straight lines passing by the points A and B:

a) $A(3, 0)$, $B(5, 0)$ b) $A(-2, -4)$, $B(2, -3)$ c) $A(0, -3)$, $B(3, 0)$ d) $A(0, -5)$, $B(-3, 1)$

21. Decide which of the following functions corresponds to the given graph:

a) $f(x) = \begin{cases} 2x+5 & if\ -3 \le x \le -1 \\ x+5 & if\ 0 \le x < 3 \\ 2x & if\ 3 \le x \le 8 \end{cases}$ b) $g(x) = \begin{cases} 2x+5 & if\ -3 \le x < 0 \\ 5-x & if\ 0 \le x < 3 \\ 2 & if\ 3 \le x \le 8 \end{cases}$

c) $h(x) = \begin{cases} 2 & if\ -3 < x < 0 \\ -1 & if\ 0 < x < 3 \\ 0 & if\ 3 < x < 8 \end{cases}$

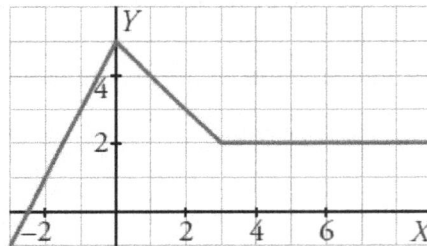

22. Represent the following functions:

a) $y = \begin{cases} 2x & if\ x \le -1 \\ -2 & if\ -1 < x \le 3 \\ x-5 & if\ x > 3 \end{cases}$ b) $y = \begin{cases} -3 & if\ x < 0 \\ 2x+1 & if\ x \ge 0 \end{cases}$ c) $y = \begin{cases} -x+3 & if\ x < 1 \\ 2 & if\ 1 \le x < 2 \\ x & if\ x \ge 2 \end{cases}$

23. Define the function whose graph is given below:

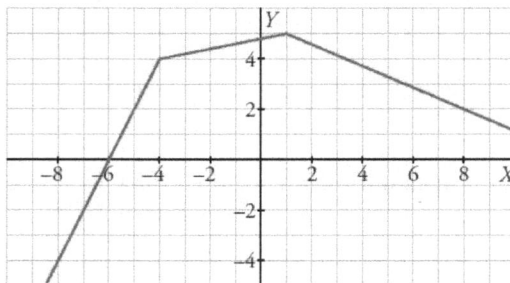

Exercises

94

Exercises

24. Calculate the values of a, b and c, for the following quadratic functions:

a) $y = 3x^2 - x + 5$ b) $y = 5x - x^2$ c) $y = x^2 + x - 8$

d) $2y = 6x - x^2 + 4$

25. Given the quadratic function $y = x^2 - 2x - 3$, represent its curve by calculating the following points by which it passes:

$y = x^2 - 2x - 3$

x	y
1	
0	
2	
-1	
3	

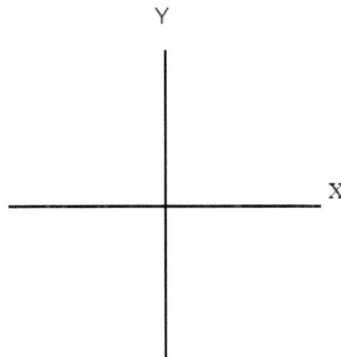

What kind of graph have you obtained?

Quadratic functions are those whose algebraic expression is $y = ax^2 + bx + c$. Its graph is a parable, whose vertex is located at the abscise: $x_v = \dfrac{-b}{2a}$

IMPORTANT: The parable will be **open up** if a > 0 and **open down** if a < 0.

Exercises

26. Find out the vertex of the following parables. a) $y = 3x^2 - 12x + 1$ b) $2y - 3x + x^2 = 5 - x - x^2$

Remember one point has two coordinates (x_v, y_v).

27. Join each quadratic function to its curve:

a) $y = x^2$

b) $y = (x - 3)^2$

c) $y = x^2 - 3$

d) $y = x^2 - 6x + 6$

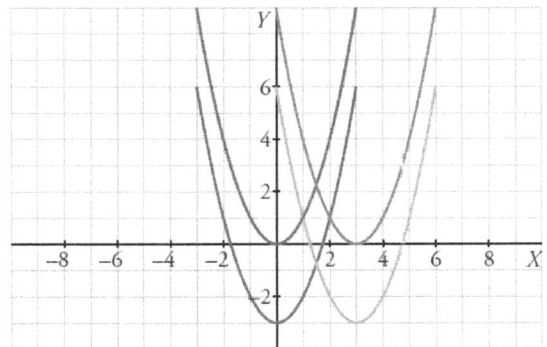

28. Represent the curve of the following parables, finding out he vertex, some close points and the intersection points with the axes:

a) $y = (x + 4)^2$ b) $y = x^2 + 2x$ c) $y = -3x^2 + 6x - 3$ d) $y = -x^2 + 5$

29. Locate the vertex (abscise and ordinated) of the following parables and indicate if it is a maximum or a minimum:

a) $y = x^2 - 5$ b) $y = 3 - x^2$ c) $y = -2x^2 - 4x + 3$

d) $y = 3x^2 - 6x$ e) $y = 5x^2 + 20x + 20$ f) $y = -x^2 + 5x - \dfrac{3}{2}$

Example: Solve, graphically and algebraically, the following system: $\begin{cases} y = x^2 + 2x - 3 \\ y = 1 - x \end{cases}$

Solution:

Graphically:

We represent the curve of the parable $y = x^2 + 2x - 3$

- Vertex: $x_v = \dfrac{-b}{2a} = \dfrac{-2}{2} = -1$ $\rightarrow$ $y_v = 1 - 2 - 3 = -4$ $\Rightarrow$ $V(-1, -4)$

- Intersections with the axes:

Y-axis $\rightarrow$ x = 0 $\rightarrow$ y = -3 $\rightarrow$ (0, -3)

X-axis $\rightarrow$ y = 0 $\rightarrow$ $x^2 + 2x - 3 = 0$ $\rightarrow$ $x = \dfrac{-2 \pm \sqrt{4 + 12}}{2} = \dfrac{-2 \pm 4}{2} \begin{cases} x_1 = 1 & (1,0) \\ x_2 = -3 & (-3,0) \end{cases}$

- Points close to the vertex:

X	-4	-2	-1	0	2
Y	5	-3	-4	-3	5

We represent the straight line $y = 1 - x$

x	1	0
y	0	1

Notice both graphs intersect at the points (-4, 5) and (1, 0).

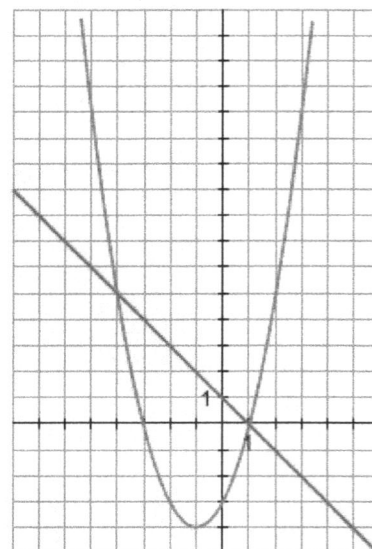

96

<u>Algebraically</u>: As y is already isolated at both expressions, we can equalize them:

$$\begin{cases} y = x^2 + 2x - 3 \\ y = 1 - x \end{cases}$$

$$x^2 + 2x - 3 = 1 - x \;\rightarrow\; x^2 + 3x - 4 = 0 \;\Rightarrow\; x = \frac{-3 \pm \sqrt{9+16}}{2} = \frac{-3 \pm 5}{2} \begin{cases} x_1 = -4 \\ x_2 = 1 \end{cases}$$

$$\begin{cases} x_1 = -4 \;\rightarrow\; y = 1 + 4 = 5 \\ x_2 = 1 \;\rightarrow\; y = 0 \end{cases} \quad\rightarrow\quad \text{So, solutions are } \boxed{\text{(-4, 5) and (1, 0)}}.$$

Exercises

30. Solve, graphically and algebraically, the following systems:

a) $\begin{cases} y = x^2 - 4x + 5 \\ x - y - 3 = 0 \end{cases}$
b) $\begin{cases} y = 2x^2 + 8x - 11 \\ y + 3 = 0 \end{cases}$

31. Check, graphically and algebraically, that the following systems do not have solution:

a) $\begin{cases} y = \dfrac{1}{2}x^2 - x - \dfrac{3}{2} \\ y = \dfrac{x}{2} - 3 \end{cases}$
b) $\begin{cases} y = \dfrac{1}{x-1} \\ y = -x + 1 \end{cases}$

32. Represent the following functions:

a) $f(x) = \begin{cases} -1 - x & if \; x < -1 \\ 1 - x^2 & if \; -1 \le x \le 1 \\ x - 1 & if \; x > 1 \end{cases}$
b) $f(x) = \begin{cases} x^2 & if \; x < 0 \\ -x^2 & if \; x \ge 0 \end{cases}$

---

8. Inverse proportionality functions $y = K/x$ and $y = (ax + b)/(cx + d)$

Exercises

33. Represent the curve of the function $y = \dfrac{2}{x}$ by completing the following values chart:

x	-4	-3	-2	-1	0	1	2	3	4
f(x)									

Function $\mathbf{f(x)} = \dfrac{\mathbf{k}}{\mathbf{x}}$, is named ***proportionality function***. Its curve is an equilateral hyperbole and it depends on the sign of k.

Its curve approaches to both coordinated axes.

34. Represent the curve of the function $y = \dfrac{-2}{x}$ by completing the following values chart:

x	-4	-3	-2	-1	0	1	2	3	4
f(x)									

The curve of the function $f(x) = \dfrac{a\,x + b}{c\,x + d}$ is similar to $f(x) = \dfrac{K}{x}$. The only difference is that:

- It approaches to the vertical line $x = \dfrac{-d}{c}$

- It approaches to the horizontal line $y = \dfrac{a}{c}$

---

35. Represent the curve of the function $y = \dfrac{2x+1}{x-3}$ by completing the following values chart:

x	4	5	6	7	8	9	10	11	3
f(x)									

x	2	1	0	-1	-2	-3	-5	-7	-9
f(x)									

36. Do the same for the function $y = \dfrac{x-1}{x-2}$.

37. Represent the curve of the following functions, studing, first of all, the straight lines to which

they approach: a) $y = \dfrac{3x-1}{x+2}$ b) $y = \dfrac{x+4}{2x+7}$ c) $y = \dfrac{3x-7}{2x+5}$ d) $y = \dfrac{x}{x+3}$

9. Logarithms. Logarithmic functions

---

38. Complete:

a) 2 raised to ………. is 8.   b) 2 raised to ………. is 32.   c) 2 raised to ………. is 512.

You have just worked with logarithms.

a) $\log_2 8 =$ b) $\log_2 32 =$ c) $\log_2 512 =$

Definition of **logarithm**: $\log_a b = c$	$\Leftrightarrow$	$a^c = b$

Logarithms are defined only for *positive* numbers (not for 0 or negative numbers).

When you use your calculator, you will see two different keys: **log** and **ln**.

Key **log** calculates $\log_{10}$ and **ln** calculates $\log_e$

Ln is named **neperian logarithm** because of its discoverer, John Neper.

Exercises

39. By using the definition of logarithm, calculate the following logarithms:

a) $\log_2 64$

b) $\log_2 16$

c) $\log_2 \dfrac{1}{4}$

d) $\log_2 \sqrt{2}$

e) $\log_3 81$

f) $\log_3 \dfrac{1}{3}$

g) $\log_3 \sqrt{3}$

h) $\log_4 16$

40. Calculate the base of the following logarithms:

a) $\log_b 10\,000 = 2$

b) $\log_b 125 = 3$

c) $\log_b 4 = -1$

d) $\log_b 3 = \dfrac{1}{2}$

41. Solve the following logarithmic equations:

a) $5^{2x^2+1} = 125$

b) $\log_3(5x-3) = 3$

c) $2^{2x-6} = 0{,}25^{x-1}$

d) $\log_5(2x^2 - x) =$

e) $\sqrt[5]{49} = 7^{x^2+\frac{6}{25}}$

f) $\log_2(x-1) = 2$

g) $3^{3x-1} = 9^{x+6}$

h) $\log_2(x^2-5x+8) = 2$

i) $4^{x^2-8x} = 1$

j) $\log(11x-1) = -1$

42. Represent the curve of the function $y = \log_2 x$ by completing the following values chart:

x	1/4	1/2	1	2	4
f(x)					

43. Represent the curve of the function $y = \log_{1/2} x$ by completing the following values chart:

x	1/4	1/2	1	2	4
f(x)					

Review exercises

1. Curve of the function $f(x) = \dfrac{x^3}{x^2-1}$ is given below. Determine its:

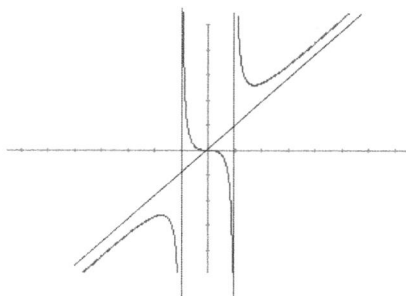

a) Intersections with the axes
b) Maximums and minimums
c) Increasing and decreasing intervals

2. Curve of the function $f(x) = \dfrac{x-1}{x+2}$ is given below. Determine its:

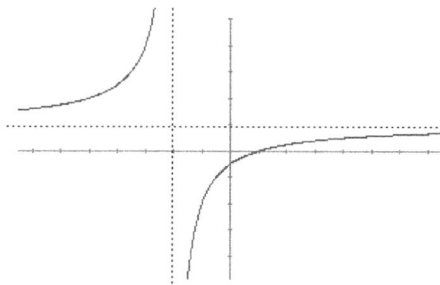

a) Intersections with the axes
b) Maximums and minimums
c) Increasing and decreasing intervals

3. Work out the most important elements of the function $f(x) = \dfrac{x^2-4}{x-3}$, whose curve is given below.

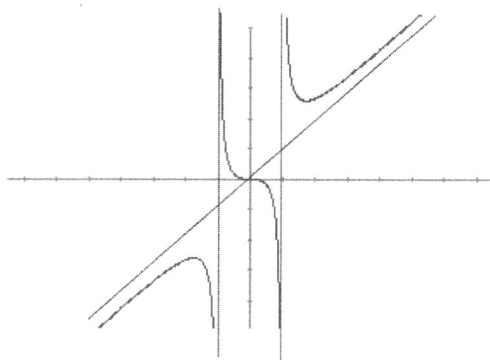

a) Intersections with the axes
b) Maximums and minimums
c) Increasing and decreasing intervals

4. Observe the curve of a function and calculate the average rate of change at the following intervals: [0, 4], [0, 5], [5, 7], [0, 7], [– 4, 0] and [–4, –2].

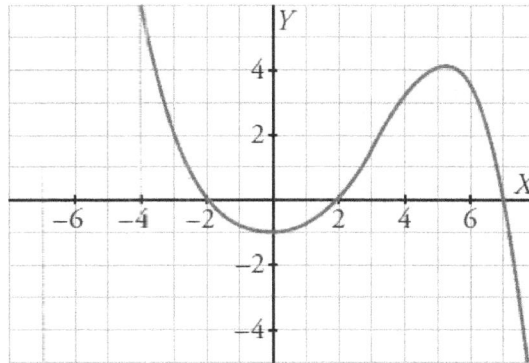

5. Calculate the average rate of change of the function: $y = 3x^3 + 9x^2 - 3x - 9$ at the following intervals: [–2, 0], [–1, 0], [–3, –1], [0, 1].

6. Location of a particle is given by the function: $s = (t^4 - 8t^3 + 18t^2)$. Calculate the average rate of this particle at the intervals: [2, 4], [1, 2], [1, 3], [2, 3].

7. Calculate *a, b* and *c* so that points A(–12, a), B(3/4, b) and C(0, c) belong to the curve of the function $y = 3x^2 - x + 3$.

8. The following curves correspond to discontinuous functions. Indicate what type of discontinuity they have.

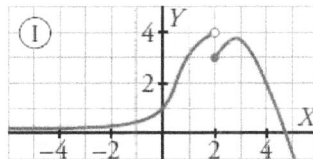

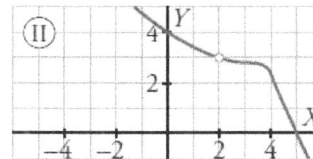

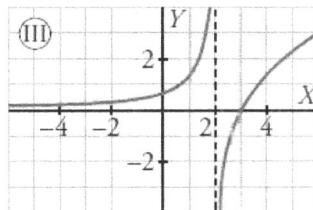

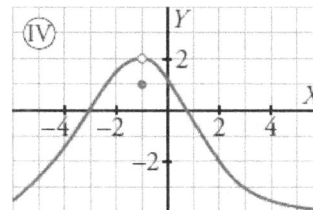

Affine functions

9. Find out the slope, y-intercept and the intersection points with the axes of the line $5x - 6y + 2 = 0$ and represent it.

10. Represent the graph of the following affine functions:

a) $y = -\dfrac{2}{5}x + 2$

b) $y = -\dfrac{3}{2}$

c) $y = \dfrac{5}{3}x$

11. Decide which of the following straight lines are parallel and represent them:

a) $y = \dfrac{x+5}{2}$

b) $y = -\dfrac{1}{2}$

c) $2x + 5y = 3$

d) $2y - x + 3 = 0$

12. Work out the equation of the straight line that passes by the points $A(1, -3)$ and $B(5, 1)$. What is its y-intercept?

13. Work out the equations of the following straight lines:

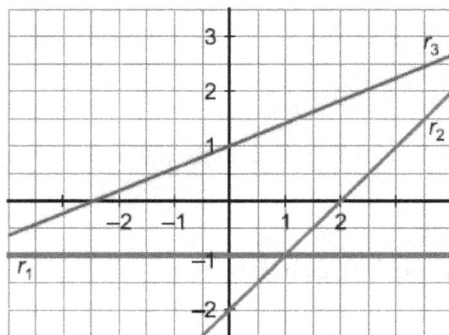

14. Work out the equation of the line that passesby the medium point of the segment AB, where $A(-1, 3)$ and $B(5, 2)$ and it is parallel to the straight line $7x - 2y + 1 = 0$.

15. Work out the equation of the following straight line:

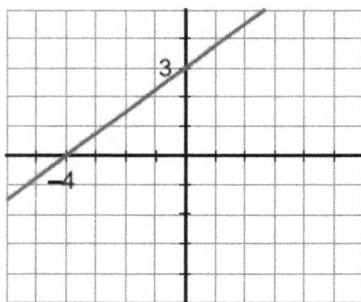

Parables

16. Represent the curve sof the following parables:

a) $y = \dfrac{1}{2}x^2 - x - \dfrac{3}{2}$

b) $y = \dfrac{1}{4}x^2 - 2x + 4$

c) $y = 2x^2 - x - 3$

d) $y = -25x^2 + 75x$

e) $y = -x^2 + 2x - 1$

17. Work the algebraic expressions of the following parables:

a)

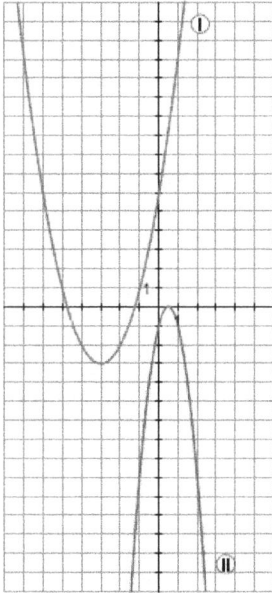

b)

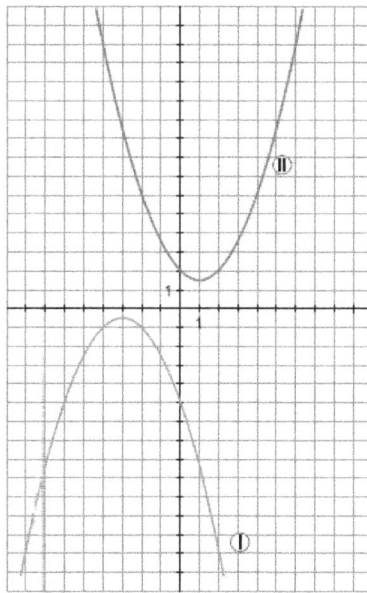

c)

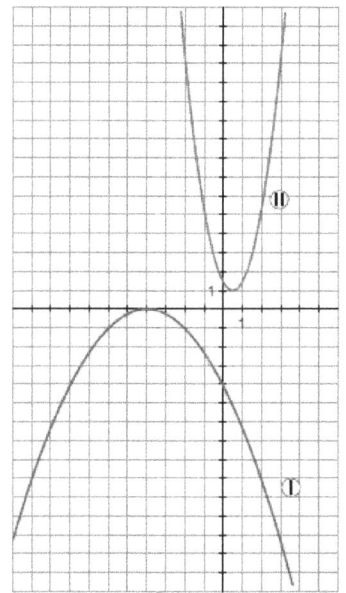

18. Join each quadratic function to its curve:

a) $y = (x - 5)^2$

b) $y = -2x^2 + 8x - 1$

c) $y = -4x^2 + 4$

d) $y = x^2 - 8x + 7$

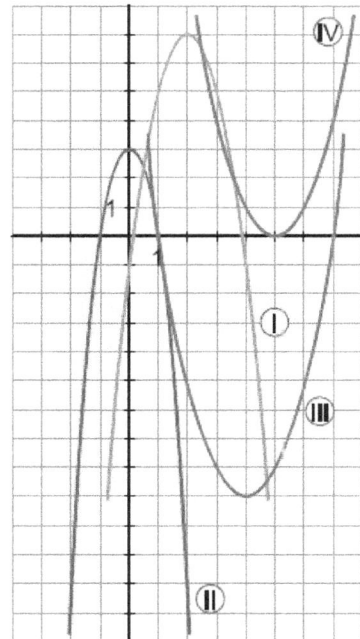

19. Join each quadratic function to its curve:

a) $y = -2x^2 + 8$

b) $y = x^2 - 3x - 10$

c) $y = -(x - 2)^2$

d) $y = 2x^2 - 3x + 1$

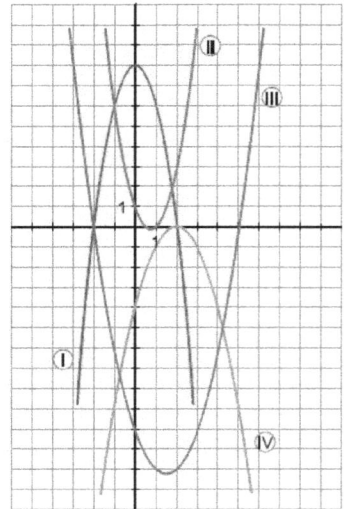

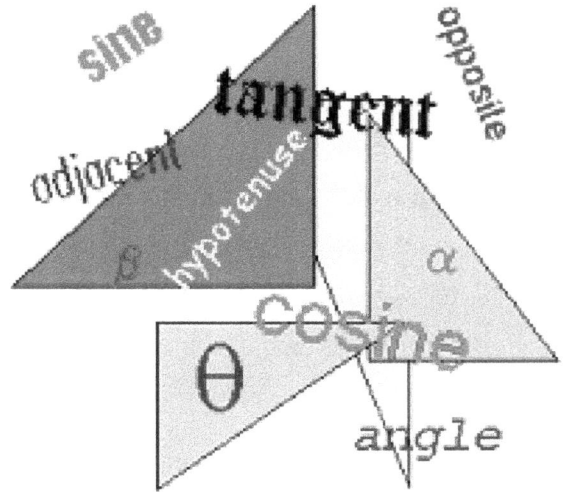

Unit 5.- Trigonometry

1. Degrees and radians

Angles are measured using two different systems of units, radians and degrees, some of whose main relationships are given.

$$360° = 2\pi\,rad \qquad\qquad 180° = \pi\,rad$$

$$90° = \frac{\pi}{2}\,rad$$

Example: Transform the following angles: a) 30°   b) $\dfrac{3\pi}{2}\,rad$

Solution:

a) $30° \cdot \dfrac{2\pi\,rad}{360°} = \dfrac{60\pi\,rad}{360} = \boxed{\dfrac{\pi}{60}\,rad}$

b) $\dfrac{3\pi}{2}\,rad \cdot \dfrac{360°}{2\pi\,rad} = \dfrac{1080\pi}{4\pi} = \boxed{270°}$

---

1. Transform into radians the following angles:

a) 45°	b) 60°	c) 90°	d) 120°	e) 150°	f) 180°
g) 210°	h) 225°	i) 270°	j) 300°	k) 330°	l) 360°

2. Transform into degrees the following angles:

a) $\dfrac{7\pi}{6}\,rad$ b) 2,5 rad   c) $\dfrac{7\pi}{3}\,rad$ d) $\dfrac{3\pi}{4}\,rad$ e) $\dfrac{5\pi}{6}\,rad$ f) $\dfrac{3\pi}{2}\,rad$

Exercises

2. Sine, cosine and tangent

When working in a right angled triangle, the longest side is known as the *hypotenuse*, and the other two sides are known as the *opposite* and the *adjacent*. The adjacent is the side next to a marked angle, and the opposite side is opposite this angle.

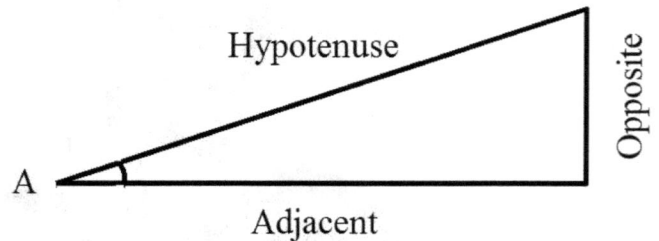

For a right angled triangle, the *sine*, *cosine* and *tangent* of the angle A are defined as:

$$Sin\,A = \frac{opposite}{hypotenuse} \qquad Cos\,A = \frac{adjacent}{hypotenuse} \qquad tg\,A = \frac{opposite}{adjacent}$$

Example: Write down the values of sinA , cosA and tanA for the triangle shown. Then use a calculator to find the angle in each case.

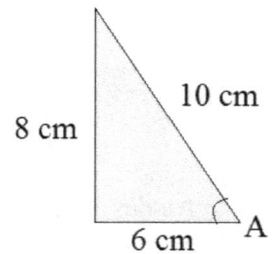

Solution:

$$Sin\,A = \frac{8}{10} = 0,8 \qquad Cos\,A = \frac{6}{10} = 0,6 \qquad tg\,A = \frac{8}{6} = 1,33...$$

- SinA = 0,8. With you calculator, press the following keys:

  SHIFT | Sin | 0 | · | 8 | = , and look at the screen: 53,13.....

  If you need this angle in degrees, then push ° ´ ´´ , and you will have **53° 7´ 48.37``**.

- CosA = 0,6. With you calculator, press the following keys:

  SHIFT | Cos | 0 | · | 6 | = , and look at the screen: 53,13.....

  If you need this angle in degrees, then push ° ´ ´´ , and you will have **53° 7´ 48.37``**.

- tgA = 0,8. With you calculator, press the following keys:

  SHIFT | tan | 1 | · | 3 | 3 | 3 | 3 | 3 | = , and look at the screen: 53,13.....

  If you need this angle in degees, then push ° ´ ´´ , and you will have **53° 7´ 48.37``**.

3. Calculate sine, cosine and tangent of the acute angles of the following right triangles:

a)

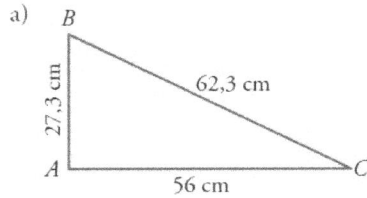

27,3 cm
62,3 cm
56 cm
B
A
C

b)

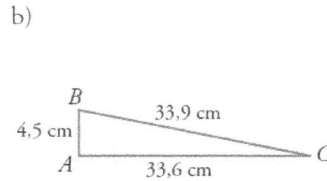

4,5 cm
33,9 cm
33,6 cm
B
A
C

c)

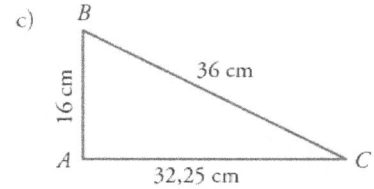

16 cm
36 cm
32,25 cm
B
A
C

4. For each triangle, write $\sin\theta$, $\cos\theta$ and $\tan\theta$ as fractions and demonstrate they fulfil the relationship: $sin^2\theta + cos^2\theta = 1$.

(a)

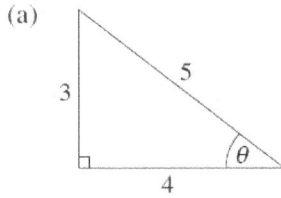

3
5
4
θ

(b)

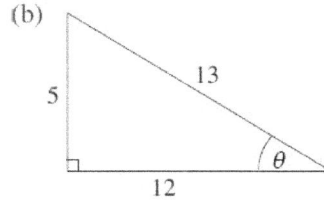

5
13
12
θ

(c)

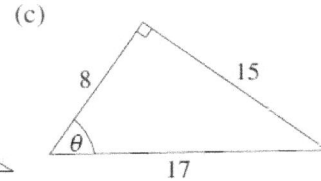

8
15
17
θ

(d)

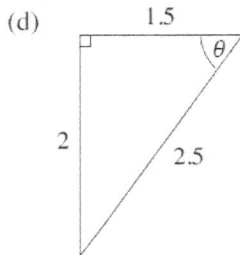

1.5
2
2.5
θ

(e)

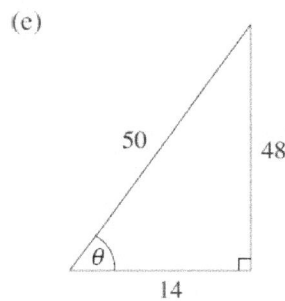

50
48
14
θ

(f)

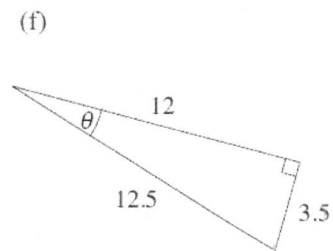

12
12.5
3.5
θ

5. Use a calculator to find the following trigonometric ratios. Give your answers correct to 3 decimal places.

a) sin30°   b) tan 75°   c) tan 52.6°   d) cos66°   e) tan33°   f) tan 45°

g) tan37°   h) sin88.2°   i) cos 45°   j) cos 48°   k) cos 46.7°   l) sin 45°

6. Use a calculator to find the angle θ in each case. Give your answers correct to 1 decimal place.

a) cosθ = 0.5   b) sinθ = 1   c) tanθ = 0.45   d) sinθ = 0.821   e) sinθ = 0.75

f) cosθ = 0.92   g) tanθ = 1   h) sinθ = 0.5   i) tanθ = 2   j) cosθ = 0.14

k) sinθ = 0.26   l) tanθ = 5.25

7. For the triangle shown, write down expressions for: a) cosθ   b) sinα   c) tanθ   d) cosα

e) sinθ   f) tanα

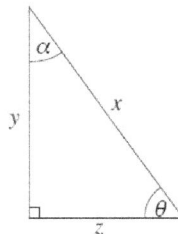

α
y
x
z
θ

8. Write the algebraic expressions of sinA and cosA and demonstrate the relationship between them: $sin^2A + cos^2A = 1$.

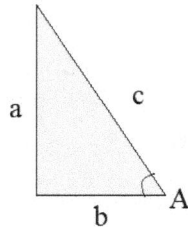

9. Knowing A is an acute angle and that $cosA = 1/5$, calculate *senA* and *tgA*.

10. Complete the following chart by using the previous relationship. A, B and C are acute angles.

	A	B	C
Sin			0,3
cos	0,25		
Tg		0,6	

12. Calculate sinA and cosA, knowing that tgA= ¾. A is an acute angle.

13. Knowing that 0° < A < 90°, complete the following chart by using the previous relationship.

	A	B
sin		0,8
cos		
tg	0,75	

14. Calculate the missing trigonometric ratios or angles, without using the calculator. 0° <A<90°:

sinA			1	
cosA	1/2			
tgA		$\sqrt{3}/2$		
A				45°

15. Calculate the missing trigonometric ratios or the angles, without using the calculator. 0° <A<90°:

sinA	$\sqrt{3}/2$			
cosA				$\sqrt{2}/2$
tgA		0		
A			30°	

Exercises

16. Calculate the missing trigonometric ratios or the angle, without using the calculator. 0° <A<90°:

sinA			1/2		
cosA					0
tgA				1	
A	0				

17. If $\sin A = 0{,}28$, calculate $cosA$ and tgA (A < 90°).

18. Find out the exact value (with radicals) of $sinA$ and tgA knowing that $\cos A = 2/3$ (A < 90°).

19. If $tgA = \sqrt{5}$, calculate $sinA$ and $cosA$ (A < 90°).

4. Another important relationships: sinA = cos(90−A)   cosA = sin(90−A)

20. Given the following right triangle, calculate sine and cosine of both acute angles and check they fulfil: $sinA = cos(90–A)$ $cosA = sin(90–A)$ Notice that $B = 90^o - A$

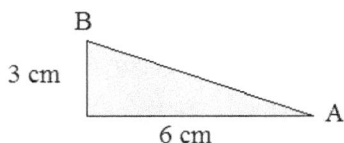

3 cm, 6 cm, B, A

21. Given the following right triangle, calculate sine and cosine of both acute angles and check they fulfil: $sinA = cos(90–A)$ $cosA = sin(90–A)$ Notice that $B = 90^o - A$

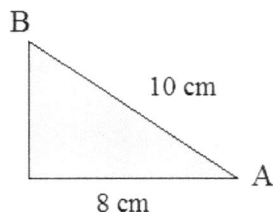

10 cm, 8 cm, B, A

22. For each triangle, write sinθ, cosθ and tanθ as fractions and demonstrate they fulfil the relationship: $sin^2\theta + cos^2\theta = 1$.

(a)

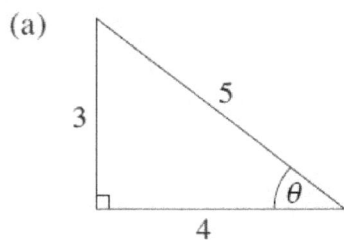

(b)

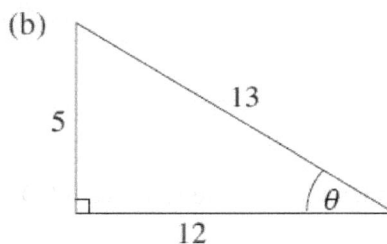

(c)

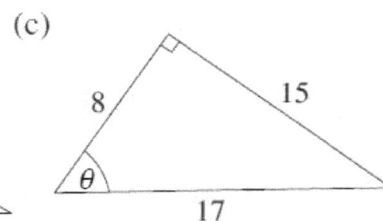

(d)

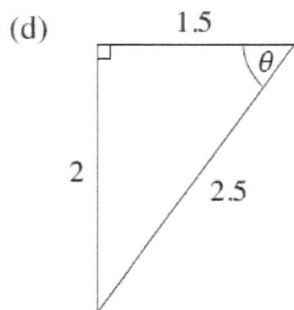

(e)

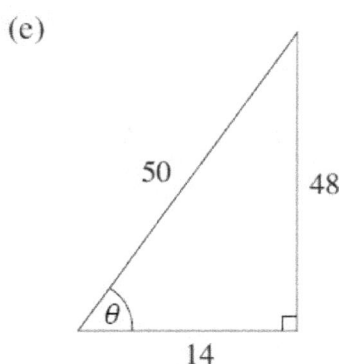

(f)

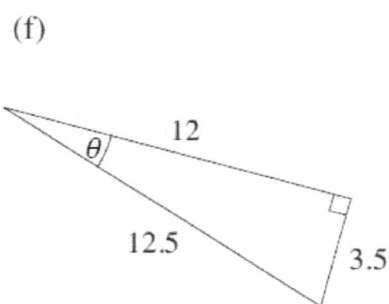

23. Use a calculator to find the following trigonometric ratios. Give your answers correct to 3 decimal places.

   a) sin30°   b) tan 75°   c) tan 52.6°   d) cos66°   e) tan33°   f) tan 45°

   g) tan37°   h) sin88.2°   i) cos 45°   j) cos 48°   k) cos 46.7°   l) sin 45°

24. Use a calculator to find the angle θ in each case. Give your answers correct to 1 decimal place.

   a) cosθ = 0.5   b) sinθ = 1   c) tanθ = 0.45   d) sinθ = 0.821   e) sinθ = 0.75

   f) cosθ = 0.92   g) tanθ = 1   h) sinθ = 0.5   i) tanθ = 2   j) cosθ = 0.14

   k) sinθ = 0.26   l) tanθ = 5.25

25. For the triangle shown, write down expressions for: a) cosθ   b) sinα   c) tanθ   d) cosα

   e) sinθ   f) tanα

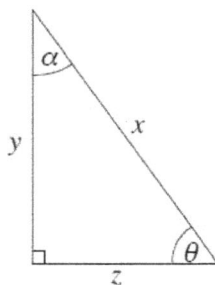

5. Calculations with right triangles

Example: Find the length of the side marked *x* in the triangle shown.

Solution: In this triangle, opposite = x and hypotenuse = 20.

Using $Sin70 = \dfrac{x}{20} = 0{,}94 \rightarrow x = 20 \cdot 0{,}94 = 18{,}8cm$

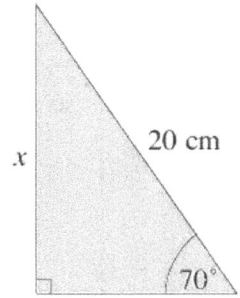

Example: Find the length of the hypotenuse, marked *x*, in the triangle.

Solution: In this triangle, opposite = 10 and hypotenuse = x.

Using $Sin28 = \dfrac{10}{x} = 0{,}47 \rightarrow x = \dfrac{10}{0{,}47} = 21{,}3cm$

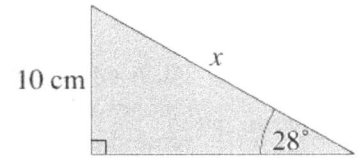

Exercises

26. The diagram below, represents one face of the roof of a house in the shape of a parallelogram *EFGH*. Angle *EFI* = 40° and *EF* = 8 m. *EI* represents a rafter placed perpendicular to *FG* such that *IG* = 5 m.

Calculate, giving your answers to 3 significant figures:

    a) the length of *FI*   b) the length of *EI*   c) the area of *EFGH*.

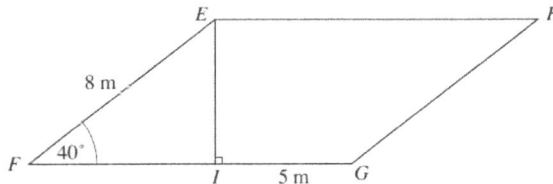

27. Find the length of the side marked *x* in each triangle:

  (a)   (b)   (c)

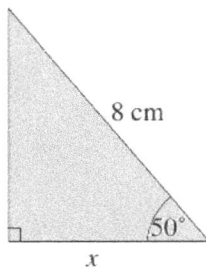
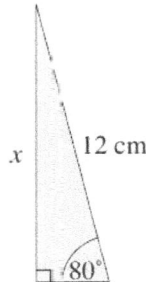
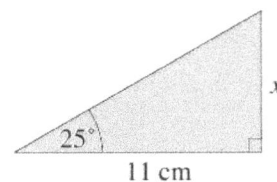

28. A ladder leans against a wall as shown in the diagram.
 a) How far is the top of the ladder from the ground?
 b) How far is the bottom of the ladder from the wall?

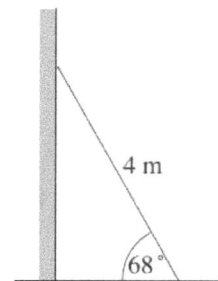

29. A guy rope is attached to a tent peg and the top of a tent pole so that the angle between the peg and the bottom of the pole is 60°.
 a) Find the height of the pole if the peg is 1 metre from the bottom of the pole.
 b) If the length of the rope is 1.4 metres, find the height of the pole.
 c) Find the distance of the peg from the base of the pole if the length of the guy rope is 2 metres.

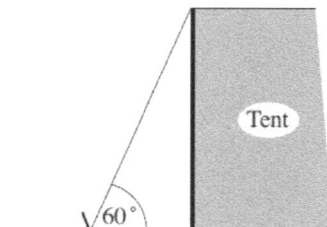

30. A child is on a swing in a park. The highest position that she reaches is as shown. Find the height of the swing seat above the ground in this position.

31. An aeroplane flies 120 km on a bearing of 210.
 a) How far south has the aeroplane flown?
 b) How far west has the aeroplane flown?

32. A kite has a string of length 60 metres. On a windy day all the string is let out an angle of between 20° and 36° with the ground. Find the minimum and maximum heights of the kite.

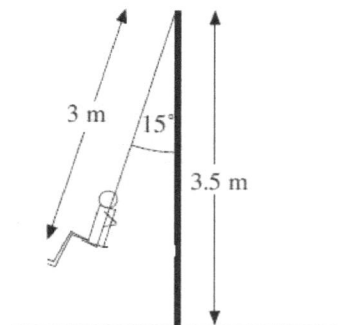

33. Find the angle marked θ in the triangle shown.

34. The diagram below shows triangle PQR. $PQ = 20$ cm, $QPR = 30°$, QS is perpendicular to PR, $SR = 9$ cm, and $SQR = x°$.   Calculate:
 a) the length of QS
 b) the size of angle x to the nearest degree.

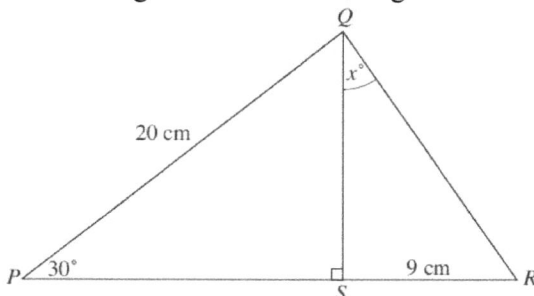

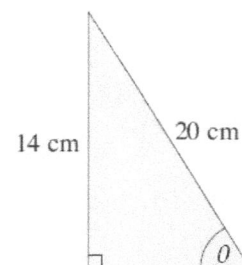

35. Find the angle in the following right triangles:

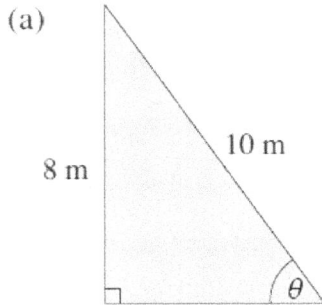

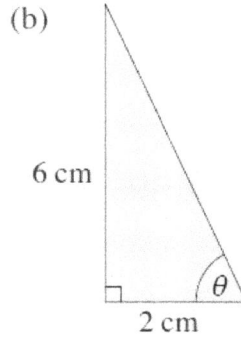

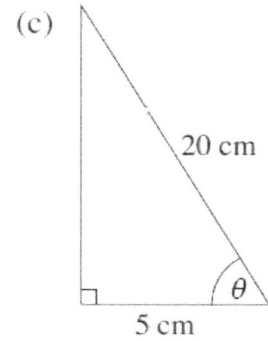

(a)

10 m

8 m

θ

(b)

6 cm

2 cm

θ

(c)

20 cm

5 cm

θ

36. A ladder leans against a wall. The length of the ladder is 4 metres and the base is 2 metres from the wall. Find the angle between the ladder and the ground.

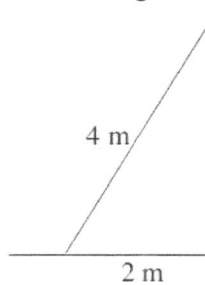

4 m

2 m

37. As cars drive up a ramp at a multi-storey car park, they go up 2 metres. The length of the ramp is 10 metres. Find the angle between the ramp and the horizontal.

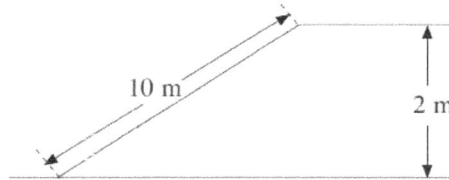

10 m

2 m

38. A soldier runs 500 metres east and then 600 metres north. If he had run directly from his starting point to his final position, what bearing should he have run on?

39. A ship is 50 km south and 70 km west of the port that it is heading for. What bearing should it sail on to reach the port?

40. Calculate the trigonometric ratios of the angles *A, C, ABD and CBD*.

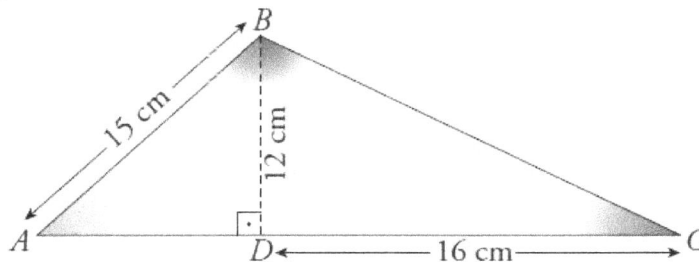

B

15 cm

12 cm

A

D

16 cm

C

6. Trigonometric ratios in the other quadrants

Till now, you have worked with trigonometric ratios in the first quadrant ($\alpha < \frac{\pi}{2}$). Now, you will learn to work with larger angles, $\alpha > \frac{\pi}{2}$.

The goniometric circumference is divided into four quadrants, called:

First	I	$0 < \alpha < \frac{\pi}{2}$	
Second	II	$\frac{\pi}{2} < \alpha < \pi$	
Third	III	$\pi < \alpha < \frac{3\pi}{2}$	
Fourth	IV	$\frac{3\pi}{2} < \alpha < 2\pi$	

Sign of the trigonometric ratios

In the goniometric circumference, angles are measured from the positive horizontal semi-axis of the first quadrant. So:

- Opposite side is located as a vertical line.

- Adjacent side is located as a horizontal line.

Remember in Maths and Physics, criteria for signs are:

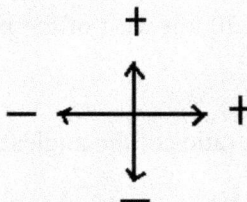

$$+$$
$$- \longleftrightarrow +$$
$$-$$

So, notice that signs for trigonometric ratios depend on the quadrant:

	I	II	III	IV
Sin	+	+	−	−
cos	+	−	−	+
Tg	+	−	+	−

Example: If $\sin \alpha = \dfrac{\sqrt{5}}{3}$ and $90° < \alpha < 180°$, what are the values of $\cos\alpha$ and $tg\alpha$?

<u>Solution:</u> Using the trigonometric relationship $\sin^2 \alpha + \cos^2 \alpha = 1$, we have:

$$\left(\frac{\sqrt{5}}{3}\right)^2 + \cos^2 \alpha = 1 \quad \rightarrow \quad \frac{5}{9} + \cos^2 \alpha = 1 \quad \rightarrow \quad \cos^2 \alpha = 1 - \frac{5}{9} = \frac{4}{9} \quad \rightarrow \quad \cos\alpha = \sqrt{\frac{4}{9}} = \frac{2}{3}$$

But, as we are at the second quadrant, $\cos\alpha < 0$, so $\boxed{\cos\alpha = -\dfrac{2}{3}}$

Now, tangent is calculated: $tg\alpha = \dfrac{\sin \alpha}{\cos\alpha} = \dfrac{\sqrt{5}/3}{-2/3} = \boxed{-\dfrac{\sqrt{5}}{2}}$ As you already knew, in quadrant II, tg < 0.

Example: Calculate $\sin\alpha$ and $\cos\alpha$ knowing that $tg\alpha = -\sqrt{5}$ and $90° < \alpha < 180°$.

<u>Solution:</u> You cannot use the trigonometric relationship $\sin^2 \alpha + \cos^2 \alpha = 1$, because you do not know values of sine or cosine. So, you must do first: $tg\alpha = \dfrac{\sin \alpha}{\cos\alpha} = -\sqrt{5} \rightarrow \sin\alpha = -\sqrt{5} \cdot \cos\alpha$ *(Equation 1)*

At this moment, you can use the trigonometric relationship $\sin^2 \alpha + \cos^2 \alpha = 1$:

$$\left(-\sqrt{5} \cdot \cos\alpha\right)^2 + \cos^2 \alpha = 1 \quad \rightarrow \quad 5\cos^2 \alpha + \cos^2 \alpha = 1 \quad \rightarrow \quad 6\cos^2 \alpha = 1 \quad \rightarrow \quad \cos\alpha = \sqrt{\frac{1}{6}} = \frac{1}{\sqrt{6}} = \frac{\sqrt{6}}{6}$$

But, as we are at the second quadrant, $\cos\alpha < 0$, so $\boxed{\cos\alpha = -\dfrac{\sqrt{6}}{6}}$

Now, *sinα* is calculated by substituting in Equation 1 $\rightarrow$ $\sin\alpha = -\sqrt{5} \cdot \cos\alpha = -\sqrt{5} \cdot \left(-\dfrac{\sqrt{6}}{6}\right) = \boxed{\dfrac{\sqrt{30}}{6}}$

As you already knew, in quadrant II, sine > 0

Exercises

41. Find out the other trigonometric ratios:

a) $sen\, \alpha = \dfrac{1}{3} \quad \alpha \in I$

b) $\cos \alpha = -\dfrac{4}{5} \quad \alpha \in III$

c) $tg\, \alpha = \dfrac{3}{4} \quad \alpha \in III$

d) $sen\, \alpha = \dfrac{3}{5} \quad \alpha \in II$

e) $tg\, \alpha = -\dfrac{5}{12} \quad \alpha \in II$

f) $\cos\alpha = \dfrac{\sqrt{5}}{5} \quad \alpha \in I$

g) $\sin \alpha = 1/2 \quad \alpha \in II$

h) $\cos x = \dfrac{3}{4} \quad x \in IV$

More trigonometric ratios

Till now, we have presented sine, cosine and tangent of an angle. There are three more trigonometric ratios, corresponding to their inverses:

Cosecant	Inverse of the sine
Secant	Inverse of the cosine
Cotangent	Inverse of the tangent

IMPORTANT: These new ratios have the same sign than their inverses.

Example: If $\cos\alpha = -\dfrac{\sqrt{7}}{4}$ and $180° < \alpha < 270°$, find out the other trigonometric ratios of this angle:

<u>Solution:</u> Using the trigonometric relationship $\sin^2\alpha + \cos^2\alpha = 1$, we have:

$$\sin^2\alpha + \left(-\frac{\sqrt{7}}{4}\right)^2 = 1 \;\rightarrow\; \sin^2\alpha + \frac{7}{16} = 1 \;\rightarrow\; \sin^2\alpha = 1 - \frac{7}{16} = \frac{9}{16} \;\rightarrow\; \sin\alpha = \sqrt{\frac{9}{16}} = \frac{3}{4}$$

But, as we are at the third quadrant, $\sin\alpha < 0$, so $\boxed{\sin\alpha = -\dfrac{3}{4}}$

Now, tangent is calculated: $tg\alpha = \dfrac{\sin\alpha}{\cos\alpha} = \dfrac{-3/4}{-\sqrt{7}/4} = \dfrac{3}{\sqrt{7}} = \boxed{\dfrac{3\sqrt{7}}{7}}$

As you already knew, in quadrant II, tg < 0.

Now, we only need their inverses:

$$\sin\alpha = -\frac{3}{4} \qquad \cos\alpha = -\frac{\sqrt{7}}{4} \qquad tg\alpha = \frac{3\sqrt{7}}{7}$$

$$\operatorname{cosec}\alpha = -\frac{4}{3} \qquad \sec\alpha = -\frac{4\sqrt{7}}{7} \qquad \cot g\alpha = \frac{\sqrt{7}}{3}$$

Exercises

42. If $\sin\alpha$ = -2/3 and α is in the third quadrant, calculate all its trigonometric ratios.

43. Calculate $\sin\alpha$, knowing that $tg\alpha$ = 3/2 and α is in the third quadrant.

44. Calculate α knowing that $\sin\alpha$ = 1/2 and $90° < \alpha < 270°$.

45. If $\cos\alpha$ = 1/3 and Л < α < 2Л, find out all its trigonometric ratios.

46. If $\sec\alpha$ = 2 and 3Л/2 < α < 2Л, calculate all its trigonometric ratios.

47. Knowing that $\cot g\alpha$ = -1/2 and that 0 < α < Л, find out all its trigonometric ratios.

48. Knowing that $\operatorname{cosec}\alpha$ = -5 and that Л < α < 3Л/2, calculate all its trigonometric ratios.

49. Find out the other trigonometric ratios:

a) $\sin\alpha = 0.6 \quad \alpha \in II$

b) $\cos\alpha = -0.6 \quad \alpha \in III$

c) $tg\alpha = -1 \quad \alpha \in II$

d) $\csc\alpha = -2 \quad \alpha \in IV$

---

7. Relationships between trigonometric ratios in different quadrants

Usually, trigonometric ratios of some angles in the first quadrant are known. In this part of the unit, you will learn to calculate trigonometric ratios of angles belonging to other quadrants.

Although there are expressions to calculate this, it is not necessary for you to leanr them by heart, as it is easier to draw a scheme. We are seeing this in several examples.

In the expressions below, α represents the angle in the first quadrant.

Angles whose difference is 90º: $B = 90° + \alpha$

$\cos \beta = \cos (90 + \alpha) = - \sin \alpha$

$\sin \beta = \sin (90 + \alpha) = \cos \alpha$

$tag \beta = tag (90 + \alpha) = - ctg \alpha$

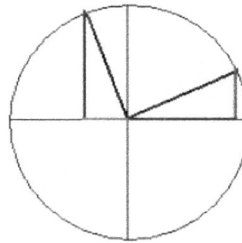

Supplementary angles: $\alpha + B = 180°$

$\cos \beta = \cos (180 - \alpha) = - \cos \alpha$

$\sin \beta = \sin (180 - \alpha) = \sin \alpha$

$tag \beta = tag (180 - \alpha) = - tg \alpha$

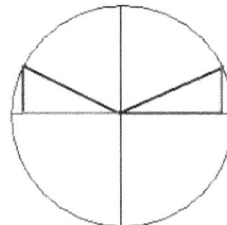

Angles whose difference is 180°: B = 180° + α

$$\cos \beta = \cos (180 + \alpha) = - \cos \alpha$$

$$\text{sen } \beta = \text{sen} (180 + \alpha) = - \text{sen } \alpha$$

$$\text{tag } \beta = \text{tag} (180 + \alpha) = \text{tg } \alpha$$

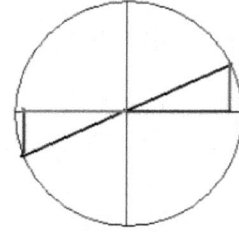

Angles whose addition is 270°: α + B = 270°

$$\cos \beta = \cos (270 - \alpha) = - \text{sen } \alpha$$

$$\text{sen } \beta = \text{sen} (270 - \alpha) = - \cos \alpha$$

$$\text{tag } \beta = \text{tag} (270 - \alpha) = \text{ctg } \alpha$$

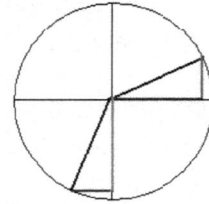

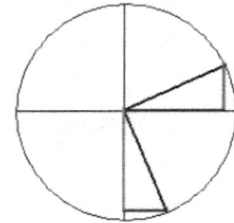

Angles whose difference is 270°: B = 270° + α

$$\cos \beta = \cos (270 + \alpha) = \text{sen } \alpha$$

$$\text{sen } \beta = \text{sen} (270 + \alpha) = - \cos \alpha$$

$$\text{tag } \beta = \text{tag} (270 + \alpha) = - \text{ctg } \alpha$$

Opposite angles: α + B = 360°   or   α + B = 0°

$$\cos (-\alpha) = \cos (360 - \alpha) = \cos \alpha$$

$$\text{sen} (-\alpha) = \text{sen} (360 - \alpha) = - \text{sen } \alpha$$

$$\text{tag} (-\alpha) = \text{tag} (360 - \alpha) = - \text{tg } \alpha$$

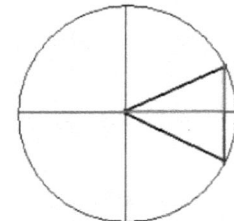

Angles whose difference is a number of whole rounds: $B = k \cdot 360° + \alpha$

$\cos \beta = \cos (\alpha + 360°k) = \cos \alpha$

$\operatorname{sen} \beta = \operatorname{sen} (\alpha + 360°k) = \operatorname{sen} \alpha$

$\operatorname{tag} \beta = \operatorname{tag} (\alpha + 360°k) = \operatorname{tag} \alpha$

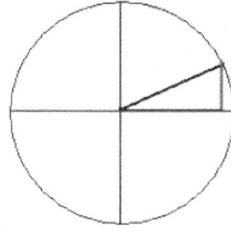

Example: Express sin 2130° as a trigonometric ratio of an angle α, so that $0 \leq \alpha \leq 90°$.

Solution: If you divide 2130: 360, you will obtain an integer quotient of 5.

$5 \cdot 360 = 1800 \quad \rightarrow \quad 2130 - 1800 = 330 \quad \rightarrow \quad \sin 2130° = \sin 330°$

Now, draw a 330° angle in the goniometric circumference:

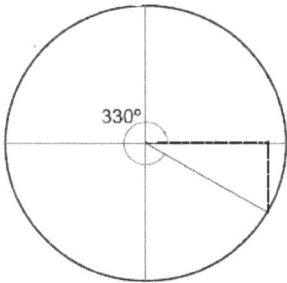

Notice the little angle under the horizontal axis. It is $360 - 330 = 30°$

Now, add a positive 30° angle in the same drawing:

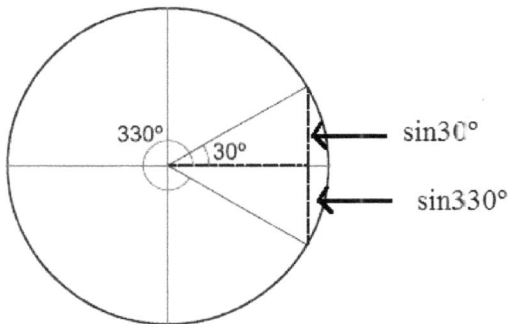

And notice they have equal sine, although with opposite signs.

So $\sin 330° = \boxed{- \sin 30°}$.

50. Transform the following trigonometric ratios into the first quadrant: $0 \leq \alpha \leq 90°$.

a) $\sin (90° - \alpha)$ b) $\sin (90° + \alpha)$ c) $\sin (\pi - \alpha)$ d) $\sin (\pi + \alpha)$

e) $\sin (- \alpha)$ f) $\sin (270° - \alpha)$ g) $\sin (3\pi/2 + \alpha)$

51. Express a trigonometric ratios of 30°:

a) $\sin 120°$ b) $\cos 120°$ c) $tg\ 150°$ d) $\sec 240°$

e) $\csc 300°$ f) $\cot g\ 420°$ g) $\sin 1110°$

52. Knowing that $\sin\alpha = 0'5$, calculate:

a) $\cos \alpha$ b) $tg\ \alpha$ c) $\cos (180 + \alpha)$ d) $\sec (90 - \alpha)$

e) $\csc (180 - \alpha)$ f) $tg (180 + \alpha)$ g) $\cos (- \alpha)$ h) $\cot (- \alpha)$

i) $\cos (90 - \alpha)$ j) $tg (90 - \alpha)$ k) $\sec (90 - \alpha)$ l) $\cot g (180 + \alpha) =$

53. Calculate, without using a calculator, the trigonometric ratios of:

a) 135°   b) 450°   c) 210°   d) −60°

54. Simplify the following expressions:

a) $sen\alpha \cdot \dfrac{1}{tg\alpha} =$ b) $sen\alpha.cos\alpha \cdot \left(tg\alpha + \dfrac{1}{tg\alpha} \right) =$ c) $sen^3 \alpha + sen \alpha . \cos^2 \alpha =$

d) $\dfrac{\cos^2 \alpha - sen^2 \alpha}{\cos^4 \alpha - sen^4 \alpha} =$ e) $\dfrac{\csc\alpha}{1 + \cot g^2 \alpha} =$ f) $\dfrac{\cos^2 \alpha}{1 - sen\alpha} =$

g) $2 \cdot \sqrt{3} \cdot \cos 30° + tg^2 60° - \cot g\dfrac{\pi}{4}$ h) $\cos^3 \alpha + \cos^2 \alpha . sen \alpha + \cos \alpha . sen^2 \alpha + sen^3 \alpha =$

55. Simplify the following expressions:

a) $\dfrac{sen^2 x.(1 + \cos x)}{1 - \cos x}$:   b) $\dfrac{\cos x}{tag x.(1 - sen x)}$

56. Check the following equality:

$$\dfrac{\sec^2 x}{\csc^2 x - \sec^2 x} + \dfrac{\cot g^2 x}{\cot g^2 x - 1} = \dfrac{1}{\cos^2 x - sen^2 x}$$

57. Solve the following trigonometric equations:

a) $sen\ x = 0$

b) $sen\ (x + \Pi/4) = \dfrac{\sqrt{3}}{2}$

c) $2\ tag\ x - 3\ cotag\ x - 1 = 0$

d) $3\ sen^2 x - 5\ sen\ x + 2 = 0$:

e) $cos^2 x - 3\ sen^2 x = 0$

f) $2\ cos\ x = 3\ tag\ x$

8. Similar figures

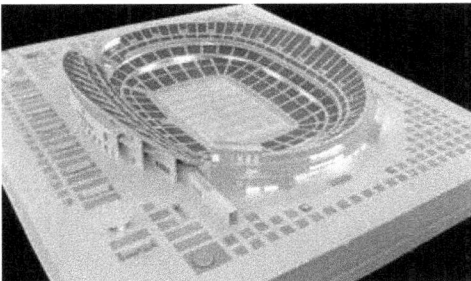

Two figures are *similar* when they have the same shape:

- Equal corresponding *angles*.
- Proportional corresponding *segments*. Proportionality ratio is named *similarity ratio*.

Some examples or similar figures are: photographs, planes, maps, three-dimensional scheme of a famous monument.

Scale of a map

The ratio of any two corresponding lengths in two similar geometric figures is called as **Scale Factor.**

So, it is the **similarity ratio** between the image and the reality.

Scale factor is denoted by **1:a** and it means that one unit in the map corresponds to "a" units in the reality.

Relationship between areas and volumes

Ratio between the areas of two similar figures is the **squared of the similarity ratio.**

Ratio between the volumes of two similar figures is the **cube of the similarity ratio.**

So, if the similarity ratio between two figures is **k**, ratio between their areas is k^2 and the ratio between their volumes is k^3.

58. In a photograph, María and Fernando measure 2,5 cm and 2,7 cm. María´s real height is 167,5 cm. What is the scale of the photograph? What is Fernando´s real height?

59. A construction company has performed a 1:90 scale model of a new mobile firm building, being square pyramid shaped. In the model, the height of the pyramid is 5.3 dm and the side of the floor is 2.4 dm. Calculate the actual volume of the building expressing the result in cubic meters.

60. Lorena takes this scheme of its kitchen to a bricklayer. What will the total area be if she decides to join both rooms?

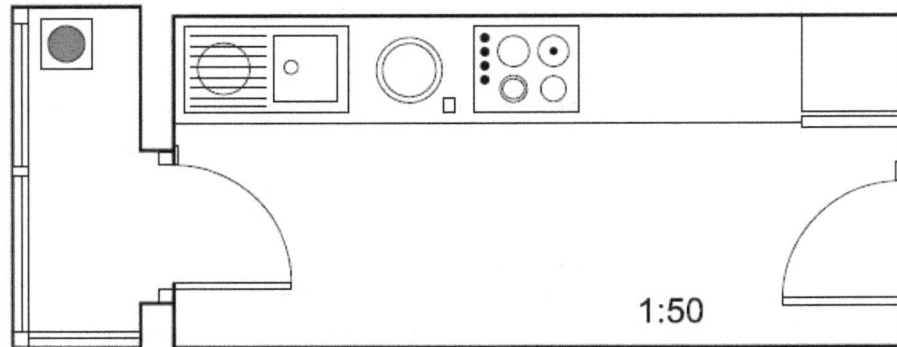

1:50

61. We want to build a frame for a photograph whose dimensions are 6 cm x 11 cm. Calculate the dimensions of the frame to make the ratio between the frame and the photograph to be 25/16.

62. On a map, whose scale is 1: 250 000, the distance between two towns is 1.3 cm.

a) What is the actual distance between the two towns?

b) What would be the distance on that map, between two other towns if they are at 15 km?

63. On a map, two towns are separated 7.5 cm. What is the scale of that map if the actual distance between the two towns is 153 km? In the same map, what would be the actual distance between two towns that are far 12.25 cm?

9. Similarity in triangles

Thales´ Theorem

Whatever pair of straight lines, if they are intersected by some parallel lines, then determined segments are proportional.

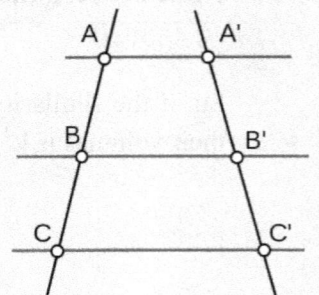

$$\frac{\overline{AB}}{\overline{BC}} = \frac{\overline{A'B'}}{\overline{B'C'}}$$

Similar triangles

If two **similar triangles** have:
- Their sides are proportional: $\dfrac{a'}{a} = \dfrac{b'}{b} = \dfrac{c'}{c} = r = similarity\ ratio$
- Their corresponding angles are equal: Â = Â' ; B = B' ; C = C'

Criteria for triangles similarity

As we are seeing, concept of similarity is very useful when solving triangles. But, first of all, we must be sure that two triangles are similar. These are some criteria you can use. You will know two triangles are similar if you can prove they fulfil, at least, one of the following conditions:

FIRST CRITERIUM: Two triangles are similar if they have two pairs of equal angles: Â = Â' ; B = B'.

SECOND CRITERIUM: Two triangles are similar if their sides are proportional: $\dfrac{a'}{a} = \dfrac{b'}{b} = \dfrac{c'}{c}$

THIRD CRITERIUM: Two triangles are similar if they have an equal angle (Â = Â') and the pair of sides that form it are proportional ($\dfrac{b'}{b} = \dfrac{c'}{c}$).

Triangles in Thales´ position

Two triangles are in **Thales' position** if:
- they have a common angle.
- sides opposite to the common angle are parallel

Two triangles in Thales´ position are **similar triangles**.

Exercises	**64.** A swimming pool is 2.3 m wide. Situating at 116 cm from the edge, from a height of 1.74 m, we note that the visual joins the edge of the pool with the bottom line. What is the deep of the pool?
	65. Calculate the height of a house knowing that at a given time of the day it produces a shadow of 3.5 m and a person who is 1.87 m has, at the same instant a shadow of 85 cm.

Similarity criterion for right triangles

Two right triangles are similar if one of their acute angles are equal.

CONSEQUENCES:

- All the triangles obtained when drawing lines being perpendicular to a side of a right triangle, are similar.

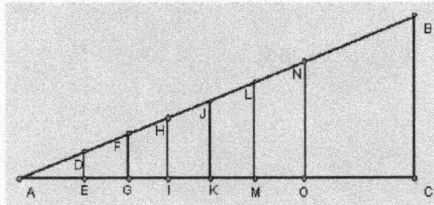

- In a right triangle, the height on the hypotenuse determines two triangles being similar to the original one.

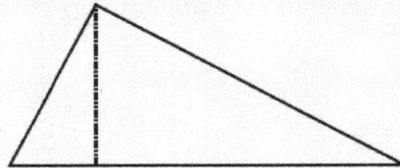

Leg´s Theorem

Squared of a leg equals the product of the hypotenuse and the projection of that leg on the hypotenuse.

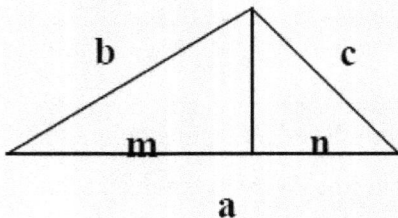

$$b^2 = a \cdot m \qquad c^2 = a \cdot n$$

Height´s Theorem

Squared of the height on the hypotenuse equals the product of the two segments determined by the intersection f the height and the hypotenuse.

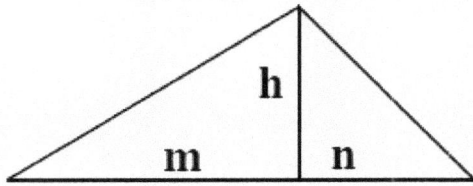

$$h^2 = m \cdot n$$

66. We want to build a bed triangle shaped garden. It is known that the height and projection of one leg on the larger side (hypotenuse) measured 15.3 m and 8.1 m. Find out its perimeter.

67. Two shops are located in the same building on the same side. Cristina is on the opposite building and wants to buy something. Calculate the distances from Cristina to both shops and from Cristina to the kiosk.

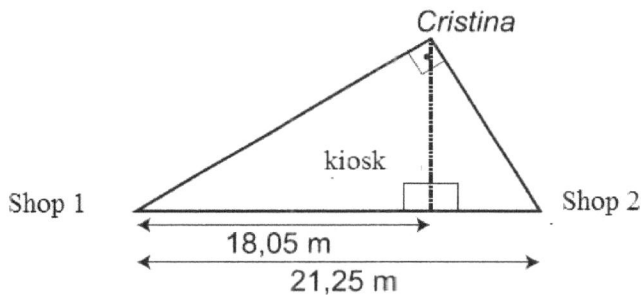

68. Antonio and Víctor houses are at the same straight Street. Every day, they go to a sport center that forms a right triangle with their houses. Observe the figure and answer:

a) What is the distance between Victor and the sport centre?

b) What i the distance between both houses?

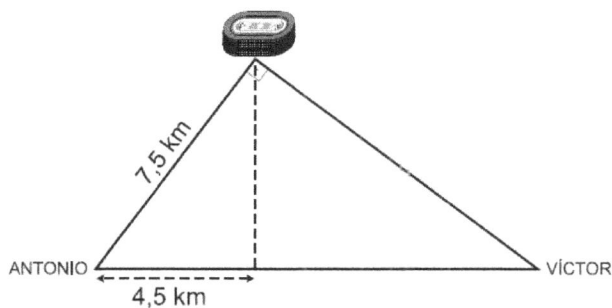

Exercises

125

69. The following figure shows the walk that a person is having. Calculate the total distance knowing that AC = 5 km and the distance from B to the shelter is 2,4 km.

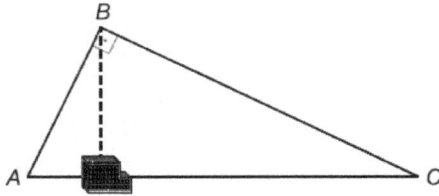

71. A boat is between two separate harbours (straight line) 6.1 km. Between both harbours there is a beach at 3.6 km from one of them. Calculate the distance between the ship and the harbours knowing that if the boat were directed towards the beach, I would move perpendicular to it. What is the distance between the ship and the beach? (NOTE: The angle of the boat with the two harbours is 90°).

72. Two parallel paths are joined by two bridges, which also intersect themselves at a point O. Observe the figure and calculate the lengths of both bridges.

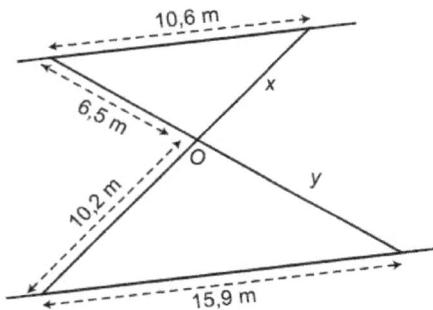

73. Between Sergio, whose height is 152 cm, and a tree, there is an oil stain on which the tree crown is reflected. Calculate the height of the tree.

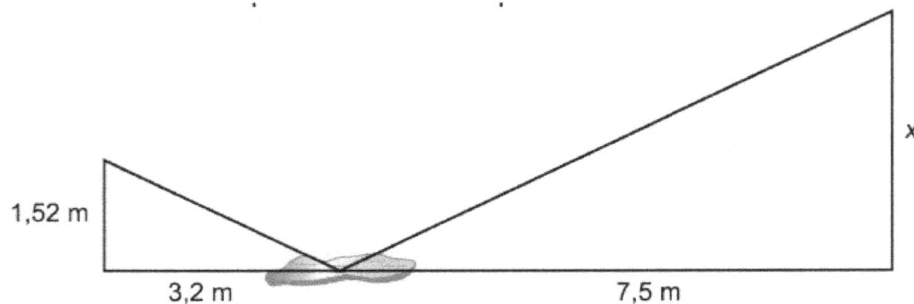

74. A tower is 100 m high. At a moment, a vertical 60 cm stick produces a 40 cm shadow. How long is the shadow of the tower at the same moment?

75. To determine the height of a mountain, Pedro, whose height is 182 cm, locates at 2,3 m from a 3,32 m tree, between himself and the mountain. Pedro is located at 138 m from the foot of the mountain. Calculate the height of the mountain.

76. An architect made a scale model 1: 100 of a cubic building for offices whose edge measures 70 m. Calculate the plant surface and the volume that the building will have on the layout.

77. The sides of two regular pentagons measure 7 cm and 5 cm. Are they similar? If so, calculate the ratio of similarity between their areas.

78. A rectangle has dimensions 3 cm x 6 cm. Find the area of another rectangle like him, knowing that the ratio of their areas is 9/4.

Review exercises

1. Calculate sine, cosine and tangent of the acute angles of the following right triangles:

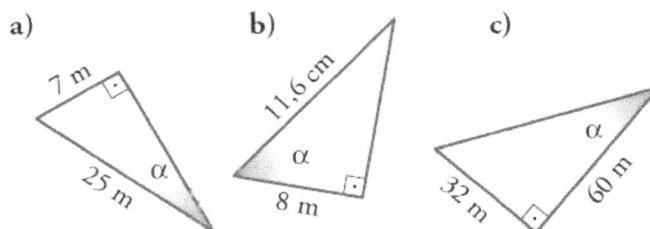

a)

7 m
25 m
α

b)

11,6 cm
α
8 m

c)

α
32 m
60 m

2. The diagram shows a slide in a play-park.

    a) Find the height of the top of the slide.

    b) Find the length of the slide.

Steps
2 m
Slide
70°
40°

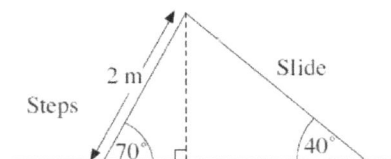

3. A snooker ball rests against the side cushion of a snooker table. It is hit so that it moves at 40° to the side of the table. How far does the ball travel before it hits the cushion on the other side of the table?

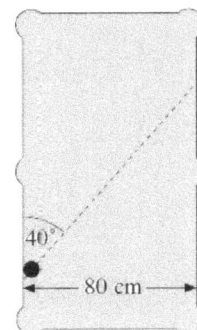

40°
80 cm

4. Find the length of the dotted line and the area of this triangle.

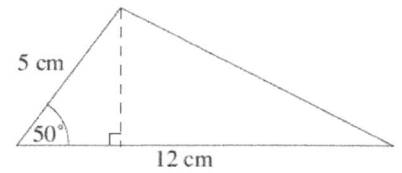

5 cm
50°
12 cm

5. A wire 18 metres long runs from the top of a pole to the ground, as shown in the diagram. The wire makes an angle of 35° with the ground. Calculate the height of the pole. Give your answer to a reasonable degree of accuracy.

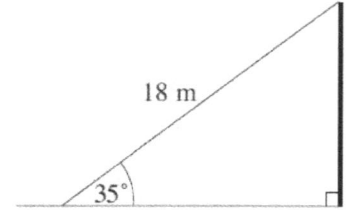

18 m
35°

6. In the figure shown, calculate

a) the length of BD.

b) the length of BC.

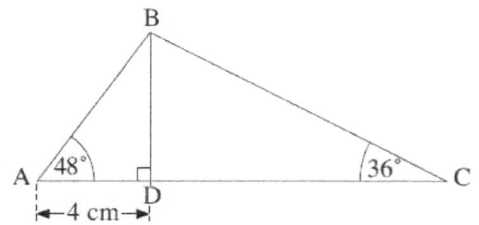

B
A 48°   36° C
D
←4 cm→

7. Find the angle in the following right triangles:

a)

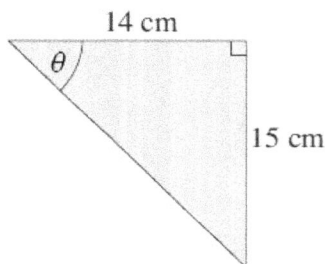

14 cm
θ
15 cm

b)

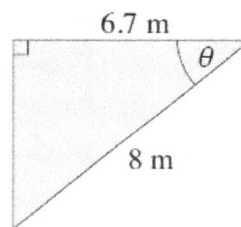

6.7 m
θ
8 m

c)

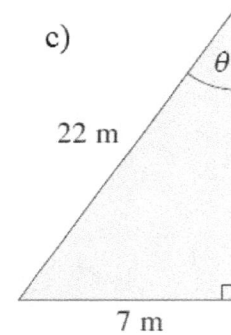

θ
22 m
7 m

8. A flag pole is fixed to a wall and supported by a rope, as shown. Find the angle between:

a) the rope and the wall,

b) the pole and the wall.

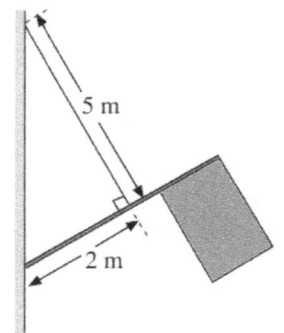

5 m
2 m

9. The mast on a yacht is supported by a number of wire ropes. One, which has a length of 15 metres, goes from the top of the mast at a height of 10 metres, to the front of the boat.

a) Find the angle between the wire rope and the mast.

b) Find the distance between the base of the mast and the front of the boat.

10. a) Calculate *x and y* in the triangle:

b) Find out the sine, cosine and tangent of the angles α and β.

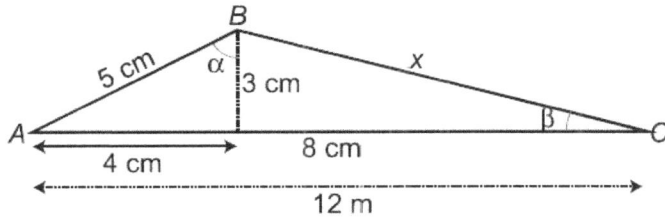

11. Find out the other trigonometric ratios:

a) $\cos\alpha = 4/5$	$0 \le \alpha \le 90°$	b) $sen\alpha = 3/5$	$0 \le \alpha \le 90°$
c) $tg\,\alpha = 3/4$	$0 \le \alpha \le 90°$	d) $\sec\alpha = 2$	$0 \le \alpha \le 90°$
e) $\cos ec\,\alpha = 2$	$0 \le \alpha \le 90°$	f) $\cot g\,\alpha = 3$	$0 \le \alpha \le 90°$
g) $\cos\alpha = 4/5$	$270° \le \alpha \le 360°$	h) $sen\alpha = 3/5$	$90° \le \alpha \le 180°$
i) $tg\,\alpha = 3/4$	$180° \le \alpha \le 270°$	j) $\cot g\,\alpha = -2$	$90° \le \alpha \le 180°$
k) $\sec\alpha = 1$	$270° \le \alpha \le 360°$	l) $\cos ec\,\alpha = -2$	$180° \le \alpha \le 270°$

12. Transform the following trigonometric ratios into the first quadrant: $0 \le \alpha \le 90°$.
a) $\sin 390°$ b) $\cos 5500°$ c) $tg 1720°$ d) $\cos ec 835°$ e) $\sec 482°$

13. Transform the following trigonometric ratios into the first quadrant: $0 \le x \le 90°$.
a) Sin (-120°)   b) Sin 2700°   c) cos (-30°)   d) Sin (270° -x)

e) Cos (180 – x)   f) Sin (180 + x)

14. Knowing that $\sin 25° = 0,42$, $\cos 25° = 0,91$ and $tg 25° = 0,47$, calculate, without using a calculator, the trigonometric ratios of 155° and 205°.

15. Calculate the trigonometric ratios of 140° and 220°, knowing that:
$\sin 40° = 0,64$; $\cos 40° = 0,77$; $tg 40° = 0,84$,

16. Calculate, with a drawing, the following trigonometric ratios:
a) $\cos(225°)$ b) $tag(120°)$ c) $\sin(1050°)$

17. Knowing that $\sec\alpha = -4$ and $0 < \alpha < \Pi$, calculate:
a) $\cos ec(\dfrac{3\pi}{2} + \alpha)$ b) $\sin(\dfrac{\pi}{2} - \alpha)$ c) $tg(360° - \alpha)$

18. Knowing that $\sin\alpha = \dfrac{2}{3}$ and $\dfrac{\pi}{2} \le \alpha \le \dfrac{3\pi}{2} < \alpha < \Pi$, calculate:
a) $\cos(\dfrac{3\pi}{2} + \alpha)$ b) $tg(\pi - \alpha)$

19. Knowing that $\cos\alpha = -\dfrac{2}{3}$ and $\pi \le \alpha \le 2\pi < \alpha < $ Л, calculate:

    a) $\cos(\dfrac{3\pi}{2} - \alpha)$ b) $\text{tg}(\pi + \alpha)$

20. Knowing that $\text{tg}\alpha = \dfrac{1}{2}$ and $\pi \le \alpha \le \dfrac{3\pi}{2} < \alpha < $ Л, calculate:

    a) $\sin(\dfrac{\pi}{2} + \alpha)$ b) $\cos(\pi + \alpha)$ c) $\text{tg}(\dfrac{\pi}{2} - \alpha)$ d) $\cot g(\pi - \alpha)$ e) $\sec(360° - \alpha)$

21. Find the height of a building that produces a shadow of 56 m at the same time that a 21 m tree produces a 24m shadow.

22. On a map, the distance between La Coruña and Lugo is 19 cm, between Santiago de Compostela and La Coruña 12 cm and between Santiago de Compostela and Lugo 20 cm. In another map, the distance between Santiago de Compostela and La Coruña is 18 cm. What are the other two distances measured in this second map?

23. On a map (scale 1: 10,000,000), the distance between two cities is 12 cm. What is the actual distance between them?

24. We have two similar isosceles triangles. About the Little one, we know that each of the equal sides is 5 cm and the unequal side 3 cm; about large one, we only know that the different side is 7 cm. How long is each of the other two sides?

25. Find the hypotenuse of a right triangle whose legs are 12 and 5 cm.

26. Knowing that in a right triangle the hypotenuse is 25 m and one leg is 7 m long, find the other leg.

27. Find the height and the area of an equilateral triangle of side 2.5 m.

28. A 3 m vertical stick produces a shadow of 2 m, what is the height of a tree if at the same time it produces a shadow of 4.5 m?

29. The lengths of the sides of a triangular field are 125 m, 75 m and 100 m. We have a drawing of the field, in which the longest side is represented by a segment of 3 cm. What are the lengths of the other two sides of the triangle in the drawing?

30. If a field is drawn using a scale of 1: 1200, what will be the actual distance that is represented by 18 cm in the drawing?

31. What is the scale used to draw a field if a 12 cm segment in the plane represents 60 m of land?

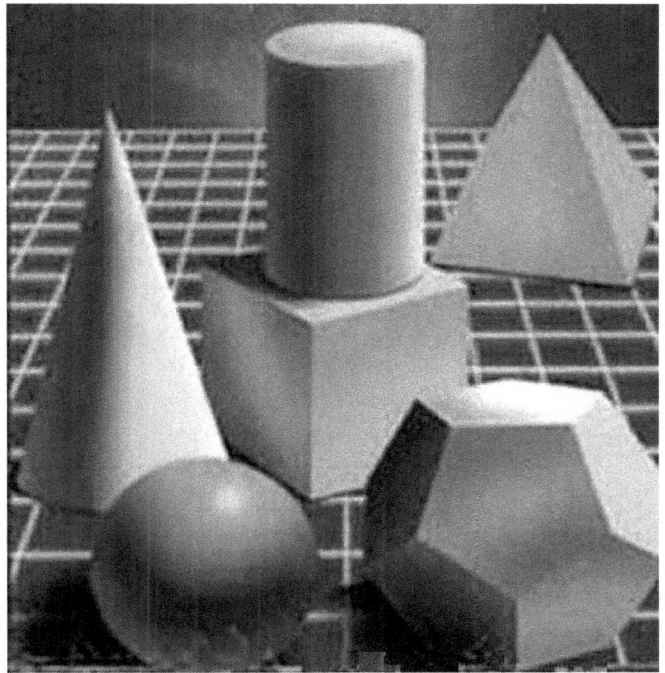

Unit 6.- Areas and volumes

1. Areas of solids

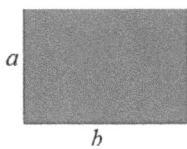

$$A = b \cdot a$$

$$A = l^2$$

$$A = b \cdot a$$

$$A = \frac{D \cdot d}{2}$$

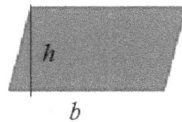
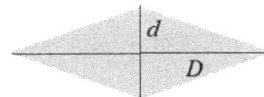

$$A = \frac{b + b'}{2} \cdot h$$

$$A = \frac{Per \cdot ap}{2}$$

$$A = \frac{b \cdot h}{2}$$

$$A = \frac{b \cdot a}{2}$$

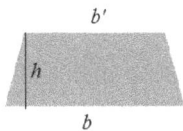
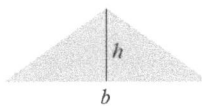
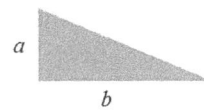

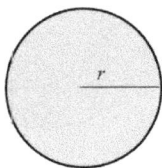
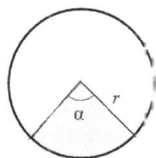
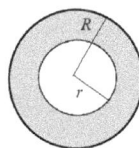
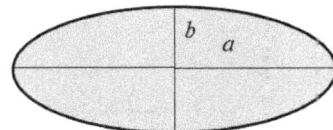

$$A = \pi \cdot r^2$$

$$A = \pi \cdot r^2 \cdot \frac{\alpha}{360}$$

$$A = \pi \cdot (R^2 - r^2)$$

$$A = \pi \cdot a \cdot b$$

131

And two important solids: Sphere and Cone:

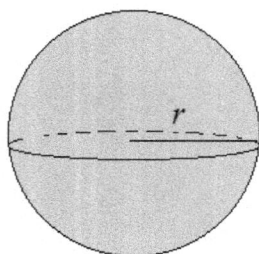

$$A = 4 \cdot \pi \cdot r^2$$

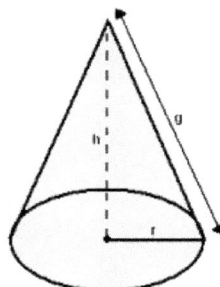

$$A = \pi \cdot r \cdot (g + r)$$

1. Calculate the area of these polyhedrons, obtained from cubes of edge 12 cm.

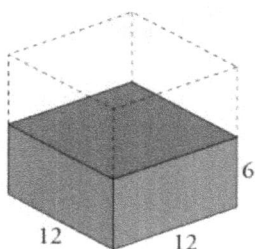

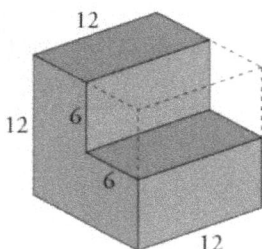

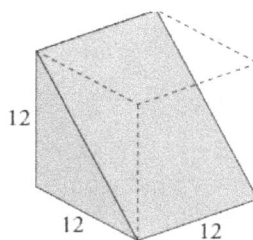

2. Calculate the area of the prism and the pyramid. In both cases, base is a regular hexagon.

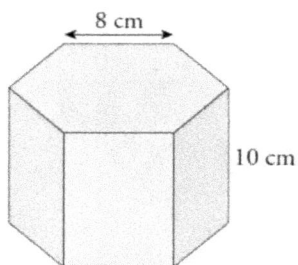

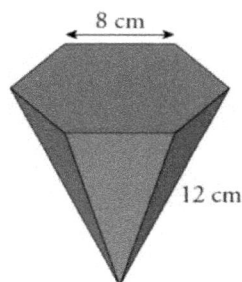

3. Calculate the area of these solids:

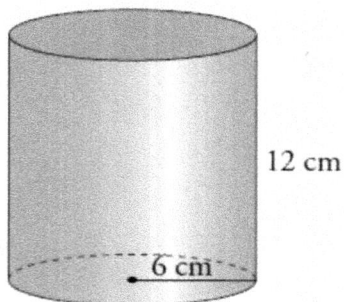

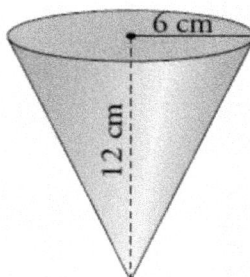

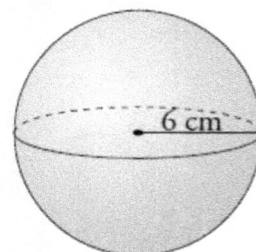

4. Calculate the area of these solids.

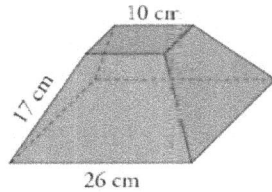

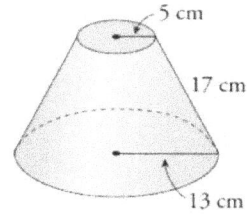

5. Calculate the areas of:

    a) A right prism whose base is a rhombus of diagonals 12 cm and 20 cm, knowing that its lateral edge measures 24 cm.

    b) A right pyramid with the same base and edge that previous prism.

2. Volume of solids

In order to organize correctly this part of the unit and make it easier to you, we are going to divide solids into three groups: right, appointed and curved solids.

- **Having two bases:** Those having two identical parallel bases. Their representatives are prism and cylinder.

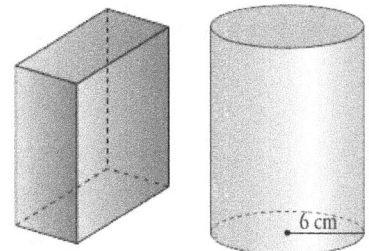

- **Having 1 base and 1 upper vertex:** Those having a base and an appointed head. Their representatives are pyramid and cone.

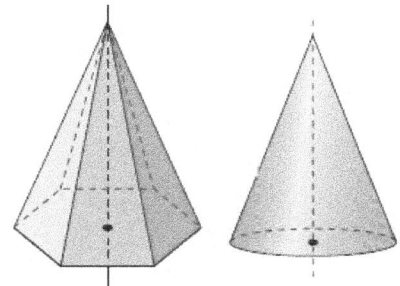

- **Curved solids:** Sphere and those derived from sphere.

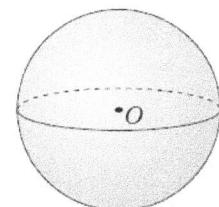

Corresponding solids: As you can see, a pyramid can be imagined into a prism, or a cone into a cylinder. We are going to name these pairs of solids, *corresponding solids*.

Now, study of solids' volume is going to be very systematical:

SOLID	VOLUME
2 bases	$V = A_{base} \cdot h$
1 base, 1 vertex	$V = \dfrac{V_{CORRESPONDING}}{3}$ or $V = \dfrac{A_{base} \cdot h}{3}$
Sphere	$V = \dfrac{4}{3} \cdot \pi \cdot R^3$

Exercises

6. Calculate the volume of these solids, obtained from prisms.

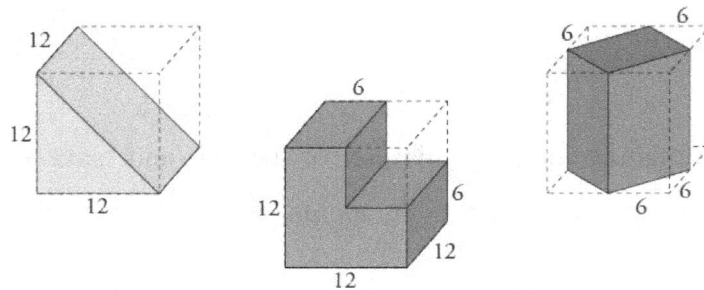

7. Calculate the volume of these solids, whose bases are regular polygons.

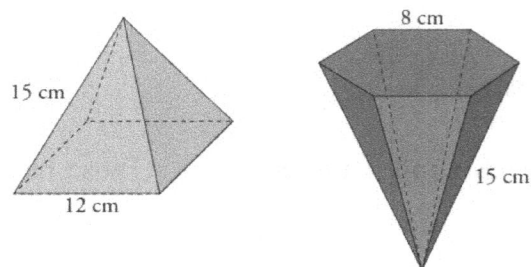

8. Calculate the volume of these solids.

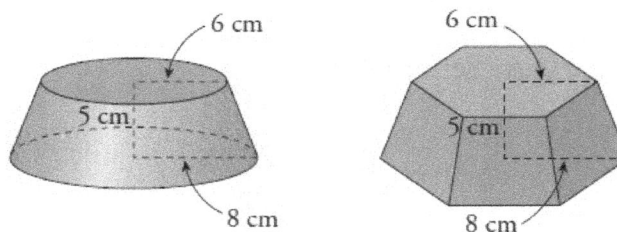

Review exercises

1. Calculate the volume existing between the cube and the cone:

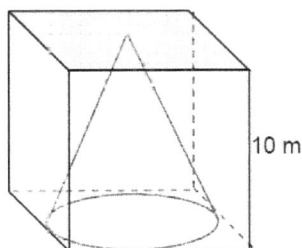

10 m

2. Calculate the area of this figure:

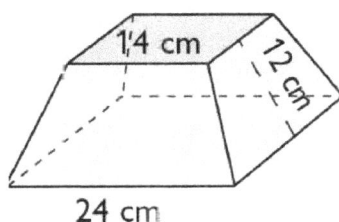

14 cm

12 cm

24 cm

3. Calculate area and volume of first figure, and volume of second one:

a)

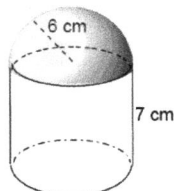

6 cm

7 cm

b)

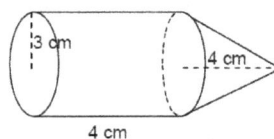

3 cm

4 cm

4 cm

4. a) Estimate the volume and the area of the Earth, knowing its radius is, approximately, 6371 km.

b) Atmosphere has a mean height of 100 km. Estimate its volume and the percent of the total volume that it represents.

5. Calculate volume of Kio Towers, in Madrid, knowing its base is a square of side 35 m and its height is 114 m.

6. We want to paint walls and ceiling of a dining-room whose floor is 12 x 7 m, and whose height is 3.5 m. This room has two doors with dimensions 1 x 2 m, and three windows with dimensions 2 x 2 m.

   a) How much area is there to paint? (Draw it).

   b) If we have painting cans for 25 m^2, how many cans do we need?

7. Calculate the volume of a Rubik's cube with edge 8 cm.

8. Calculate the volume, in ml, of a can of Coca-cola, whose height is 10.9 cm and whose diameter is 6.2 cm. (1 ml = 1 cm^3)

9. Calculate the volume of Keops' pyramid, knowing its height is 230.35 m and its base is a square of side 136.86 m.

10. A water deposit is a prism whose height is 10 m and whose volume is 4000 m^3. Calculate the side of the base, knowing it is a square.

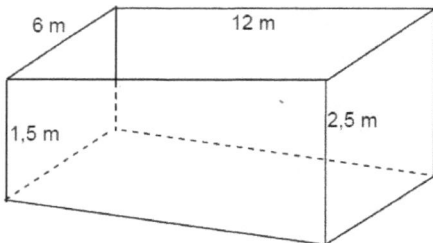

11. Calculate the volume of the shown swimming pool. (1 m^3 = 1 000 l)

12. About previous exercise, how long would it take to fill it if water comes at 0.5 l/s?

(EXERCISE 11)

6 m   12 m   2,5 m   1,5 m

13. Base diameter of a cylinder is equal to its height. Total area is 169.56 m^2. Calculate its dimensions.

14. Calculate the area of the following solids:

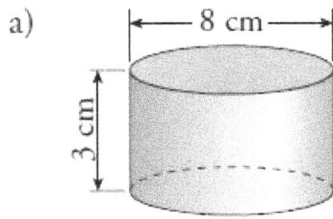

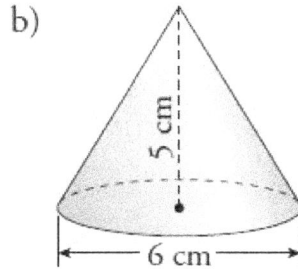

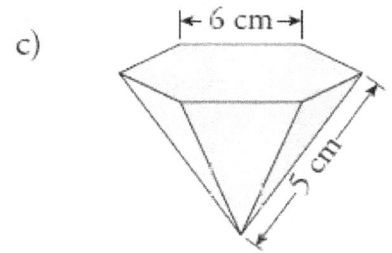

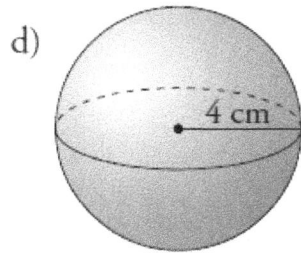

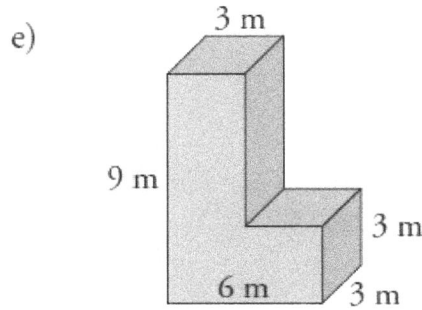

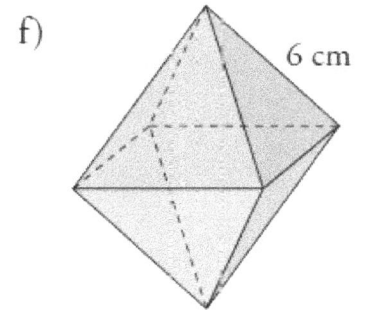

a)

8 cm

3 cm

b)

5 cm

6 cm

c)

6 cm

5 cm

d)

4 cm

e)

3 m

9 m

3 m

6 m

3 m

f)

6 cm

15. Calculate the volume of the following solids:

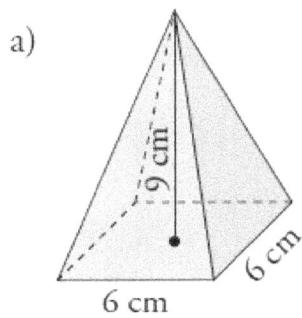

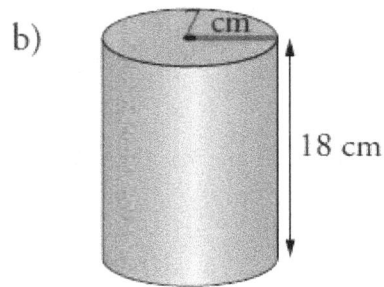

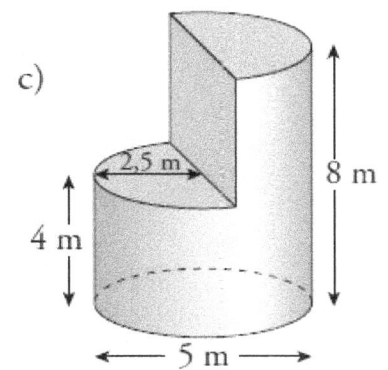

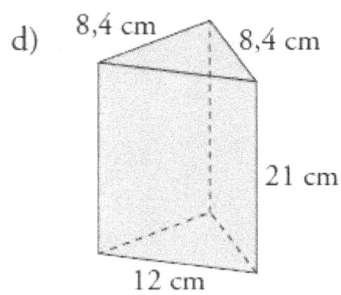

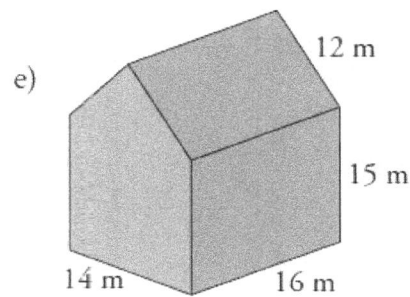

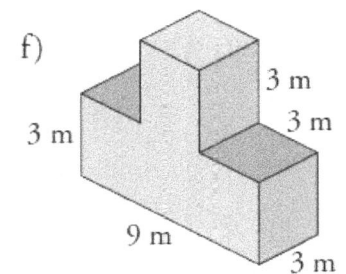

a)

9 cm

6 cm

6 cm

b)

7 cm

18 cm

c)

2,5 m

4 m

8 m

5 m

d)

8,4 cm

8,4 cm

21 cm

12 cm

e)

12 m

15 m

14 m

16 m

f)

3 m

3 m

3 m

9 m

3 m

16. Calculate the volume of following solids:

a) Regular octahedron with edge 10 cm.

b) Regular hexagonal pyramid with a lateral edge of 15 cm and basic edge of 8 cm.

c) Semi-sphere with radius 10 cm.

d) Cylinder inscribed in a right prism with a square base with side 6 cm and height 18 cm.

217. Calculate the volume of this regular tetrahedron. Clue: To calculate height H, remember that $\overline{AO} = \dfrac{2}{3}h$, where h is the height of a face.

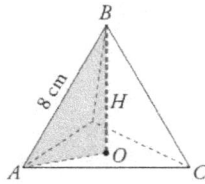

18. Calculate the volume of following solids:

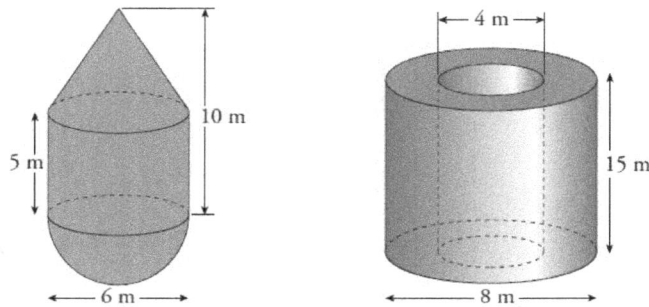

19. Calculate the volume of this solid. Look at the drawing at the right of the solid for help. Units are in meters.

20. Which of these glasses has a higher volume?

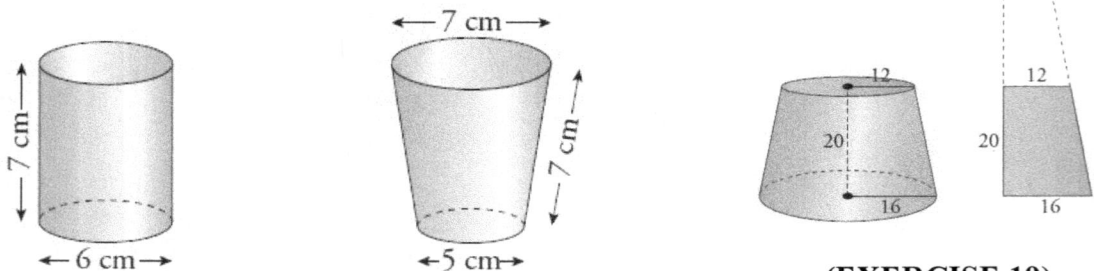

(EXERCISE 19)

138

21. What is the area of the coloured triangle?

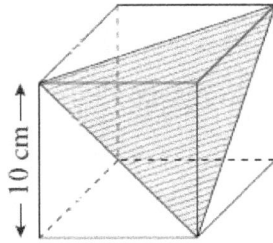

22. What is the volume of the room whose floor is shown?

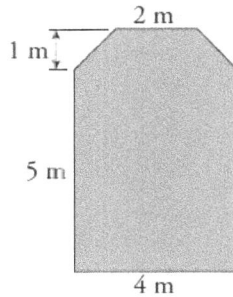

23. Calculate the volume of the solids generated when rotating these plane shapes around the indicated axis.

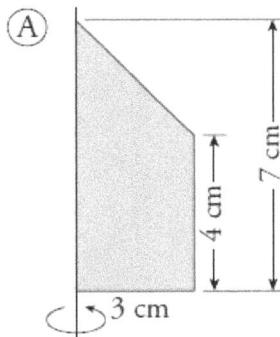

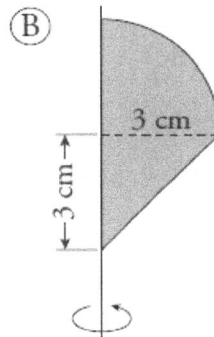

24. Three tennis balls are kept in a cylindrical box with 6.6 cm diameter. Calculate the volume of empty space.

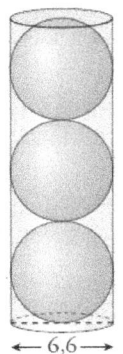

25. We put a 14 cm diameter ball in a 14 cm edge cube, initially full of water. Calculate: a) Volume of dropped water. b) Height of water remaining in the cube.

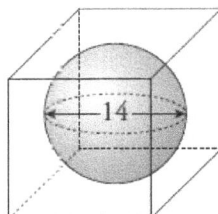

26. A isosceles right triangle has legs of 8 cm. Calcuate the volume of the solid generated when rotating around its hypotenuse.

27. We want to build a cylinder by joining parallel sides of a rectangle whose sides measure 20 and 28 cm. What sides do we have to join if we want to obtain a cylinder with a higher volume?

28. We have cut a 8 cm edge cube as shown in the figure. What is the volume of each portion?

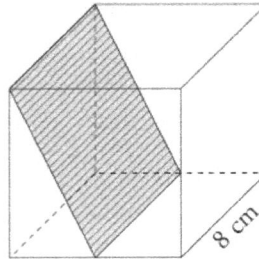

29. When opening a cone we obtain a circle sector with angle 120° and area 84.78 cm^2. Calculate the total area and the volume of the cone.

30. A cylinder and a cone have the same total area, 96π cm^2, and the same radius, 6 cm. Which of them will have a higher volume?

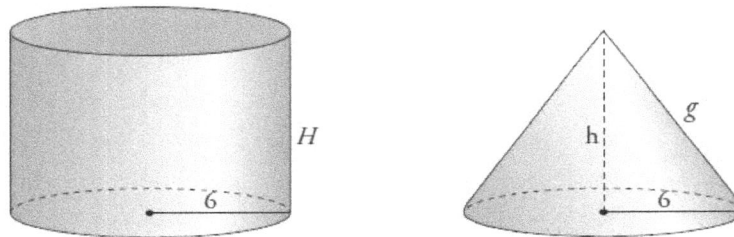

31. If in a cone we reduce into its half the base radius and keep its height, will its volume reduce into its half? And if we keep the base and reduce its height into the half?

32. A square base pyramid is cut by a horizontal plane, parallel to the base and passing by the medium point of the height. What will be relationship between volumes of higher and lower pyramids?

33. A cube and a sphere have the same area. Which of them has a higher volume? Do it by giving a value to the sphere's radius.

Unit 7.- Analytical geometry

1. Mid-point of a segment line

The coordinates of the mid-point between two other points may be found by drawing or by calculation.

Consider the line segment that joins the point A which has coordinates (2, 2) and the point B (6, 8).

- The value of the x-coordinate of the mid-point of the line segment AB is the mean value of the two x-coordinates of the end points A and B $\rightarrow$ $x_M = \dfrac{2+6}{2} = 4$

- Similarly for the y-coordinate of the mid-point, it is the mean of the y-coordinates of the end points A and B $\rightarrow$ $y_M = \dfrac{2+8}{2} = 5$ So, coordinates of the mid-point are: $\boxed{M(4, 5)}$.

Coordinates of the mid-point of the segment line AB are: $\left(\dfrac{x_A + x_B}{2}, \dfrac{y_A + y_B}{2} \right)$

Example: Find out the symmetric point A´ of A(2, 4) about the point B(-2, 6).

Solution: Coordinates of A´ are called (x, y). Notice that B is the mid-point of the segment-line AA´:

$\left(\dfrac{2+x}{2}, \dfrac{4+y}{2} \right) = (-2,6) \rightarrow \begin{cases} \dfrac{2+x}{2} = -2 \rightarrow 2+x = -4 \rightarrow x = -6 \\ \dfrac{4+y}{2} = 6 \rightarrow 4+y = 12 \rightarrow y = 8 \end{cases}$ So, coordinates of A are $\boxed{M(-6, 8)}$.

1. Calculate central point of a segment determined by points (1,4) and (5,-8).

2. A town A is located at point (-2, 5) and another one, B, at point (4,10). They decide to build a shopping centre in the central point of the road that joins them. Where will they build it?

3. Central point of a segment AB is located at point M(3,5). If A is (-4,2), where is B located?

4. Calculate the symmetric point of A(1,2) about B(-2,7). A scheme might be useful for you.

5. A frog moves by jumping on a straight line. Its first jump starts at point A(-4, -4) and finishes at point B(2,1), where it starts a new jump. Determine the point in which it will finish its fifth jump from A.

2. Equations of a straight line

Equation of a straight line is an algebraic relationship between coordinates x (abscise) and y, (ordinated) of all its points.

First of all, we will learn how to calculate the slope (or gradient) of a straight line.

Slope (or gradient) of a straight line

Slope or *gradient, m,* of a straight line is the *tangent* of the angle between the straight line and the positive part of the x-axis.

Slope of a straight line passing by points $A(x_1, y_1)$ and $B(x_2, y_2)$ is calculated as

$$m = \frac{y_2 - y_1}{x_2 - x_1}$$

Given two points with coordinates $A(x_1, y_1)$ and $B(x_2, y_2)$, we can calculate the slope of the straight line that joins them. We have only to calculate variation of *y-variable* (increasing or decreasing) when *x- variable* varies, passing from point A to B.

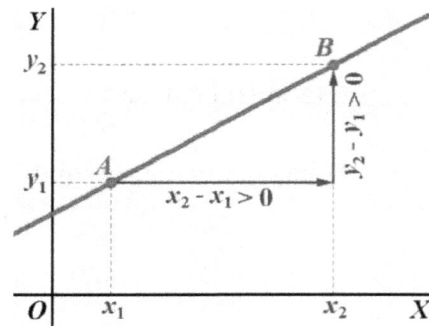

Example: Calculate the slope of the straight line passing by points with coordinates $A(-1, 1)$ y $B(1, 5)$.

Solution:

$$m = \frac{y_2 - y_1}{x_2 - x_1} = \frac{5-1}{1-(-1)} = \frac{4}{2} = 2$$

Now, look at the right triangle between the straight line and the positive part of the x-axis:

Opposite side: 5   Adjacent side: 2 $\rightarrow$ $tg\alpha = \frac{5}{2.5} = 2$

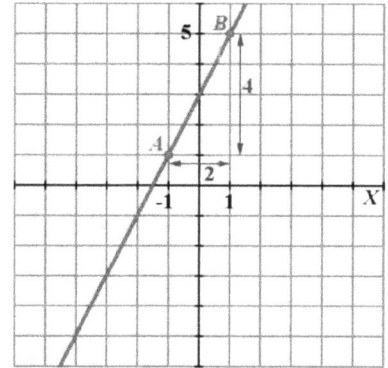

Gradient of a perpendicular straight line

Given a straight line, with slope m, the slope, **m'**, of a perpendicular direction is:

$$m' = -\frac{1}{m}$$

Example:

a) Plot the points A (1, 2) and B (4, 11), join them and calculate the gradient of AB.

b) On the same set of axes, plot the points P (5, 4) and Q (8, 3), join them and calculate the gradient of PQ.

c) With your set of rules, check these segment-lines are perpendicular and, then, check their gradients fulfil the relationship

Solution:

a) $m = \frac{11-2}{4-1} = \frac{9}{3} = 3$

b) $m = \frac{3-4}{8-5} = \frac{-1}{3} = -\frac{1}{3}$

c) Notice that $\frac{-1}{3} = -\frac{1}{3} \rightarrow m' = -\frac{1}{m}$

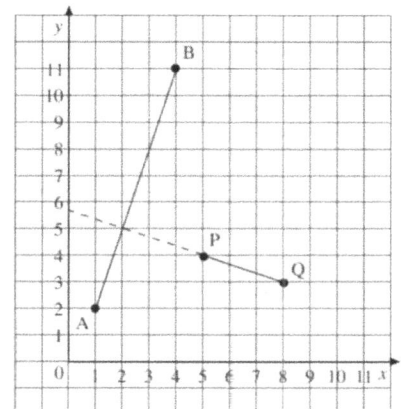

6. In each case, decide whether the lines AB and PQ are parallel, perpendicular or neither.

    a) A(4, 3) B(8, 4)   P(7, 1) Q(6, 5)

    b) A(–2, 0) B(1, 9)   P(2, 5) Q(6, 17)

    c) A(8, –5) B(11, –3)   P(1, 1) Q(–3, 7)

    d) A(3, 1) B(7, 3)   P(–3, 2) Q(1, 0)

7. The points P(–3, 1), Q(1, 2), R(0, –1) and S(–4, –2) are the vertices of a quadrilateral.

    a) Calculate the gradient of each side of the quadrilateral.

    b) Is the quadrilateral a parallelogram?

    c) Is the quadrilateral a rectangle?

8. A triangle has vertices A (3, 1), B (7, 5) and C (1, 3). Show that the triangle is a right-angled triangle.

9. The coordinates of the point A, B, C and D are A(3, 0), B(0, 1), C(1, 4) and D(4, 3). Show that ABCD is a square.

10. The points A(3, 2), B(6, 0), C(5, 4) and D(2, 6) are the vertices of a quadrilateral.

    a) Show that this is *not* a rectangle.   b) Show that this is a parallelogram.

11. The lines AB and PQ are perpendicular. The coordinates of the points are

    A(3, 2)   B(7, 4)   P(3, 7) and Q(6, q)   Determine the value of q.

Equation of the straight line passing by 2 points

As you know, the gradient of a straight line does not depend on the pair of points you consider. Let´s name X(x, y), a general point. If we are given two points of the straight line, A and B, we can calculate its gradient taking in account two pairs of points: AB and AX. These gradients are equal.

Look at the following example:

Example: Write the equation of the straight line passing by points with coordinates $A(-1, 1)$ y $B(1, 5)$.

<u>Solution:</u>

For AB pair $\rightarrow$ $m = \dfrac{5-1}{1-(-1)} = \dfrac{4}{2} = 2$ $\qquad$ For AX pair $\rightarrow$ $m = \dfrac{y-1}{x-(-1)} = \dfrac{y-1}{x+1}$

As both gradients are equal, $2 = \dfrac{y-1}{x+1}$ $\rightarrow$ $y-1 = 2x+2$ $\rightarrow$ $\boxed{y = 2x+3}$

Equation of the straight line passing by two points A and B:

$$\boxed{\dfrac{y_B - y_A}{x_B - x_A} = \dfrac{y - y_A}{x - x_A}}$$

Notice that points A and B determine vector $\overrightarrow{AB}$. This vector is called direction vector, whose coordinates are $\vec{v_d} = \overrightarrow{AB} = (x_B - x_A, y_B - y_A) = (v_x, v_y)$. So, equation below can be rewritten in a different and easier way, called **continuous equation**:

$$\dfrac{y_B - y_A}{x_B - x_A} = \dfrac{y - y_A}{x - x_A} \rightarrow \dfrac{v_y}{v_x} = \dfrac{y - y_A}{x - x_A} \quad Cross-product \rightarrow \dfrac{x - x_A}{v_x} = \dfrac{y - y_A}{v_y}$$

Continuous equation of the straight line passing by two points A and B:

$$\boxed{\dfrac{x - x_A}{v_x} = \dfrac{y - y_A}{v_y}}$$

<table>
<tr>
<td rowspan="2">Exercises</td>
<td>

12. Work out the equations of the straight line passing by the point A(-2,4) and whose direction vector is $\vec{u}\,(3,-1)$.

</td>
</tr>
<tr>
<td>

13. Work out the equations of the straight line passing by the point A(4,0) and whose direction vector is $\vec{u}\,(1,3)$.

</td>
</tr>
</table>

Exercises

14. Work out the equations of the straight line passing by the point A(-2,4) and whose direction vector is $\vec{u}$(3,-1).

15. Work out the equations of the straight line passing by the point A(4,0) and whose direction vector is $\vec{u}$(1,3).

16. Work out the equations of the straight line passing by the point A(-2,5) and whose direction vector is $\vec{u}$(4,1).

17. a) Write the equation of the straight line passing by the points (1,0) and (3,6)..

b) Write the equation of the straight line being parallel to the line $y = \frac{1}{2}x$ and passing by (4,4).

c) Locate the intersection point of both lines.

18. Workout the equation of the straight lines passing by the following points:

a) $A(-1, 0)$, $B(0, 3)$ b)$A(0, -2)$, $B(5, -2)$ c) $A(-2, 3)$, $B(4, -1)$

19. Workout the equation of the following straight lines:

a) Parallel to y = –2x + 3 and passing by (4, 5).

b) Parallel to 2x – 4y + 3 = 0 and passing by (4, 0).

c) Parallel to 3x + 2y – 6 = 0 and passing by (0, –3).

20. Workout the equation of the following straight lines:

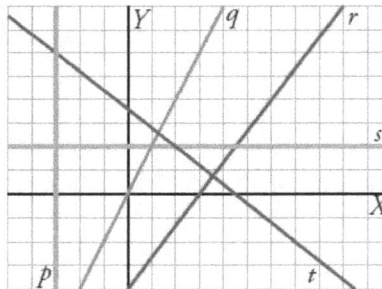

21. Write the equation of the perpendicular straight line to *r* and passing by the point *P*:

a) r : y = –2x + 3; P(–3, 2)   b) r : 3x – 2y + 1 = 0;   P(4, –1) c) r : x = 3; P(0, 4).

22. Check if the points A(18, 15) and B(–43, –5) are located in the straight line x – 3y + 27 = 0.

23. Given points A(–3, 2) and B(5, 0), write the straight lines: a) Passing by A and being perpendicular to segment-line AB; b) Passing by B and being perpendicular to segment-line AB.

146

24. Calculate *n* and *m* so that straight lines

$$r: 3x + my - 8 = 0 \qquad s: nx - 2y + 3 = 0$$

intersect at the point P(1, 5).

25. a) Write the equation of the straight line passing by (3, 2) and whose direction vector is (1, 1).

b) Write the equation of the straight line passing by (5, 2) and being parallel to X-axis.

c) Locate the intersection point of both lines.

Equation of the straight line knowing its slope and y-intercept

Equation of the straight line whose slope is *m* and that intersects the *Y-axis* at the point *(0, n)*, is:

$$y = mx + n$$

Example: Find the equation of the straight line being parallel to the one with equation $y = 4x - 2$, knowing it passes by the point with coordinates (1, 9).

Solution:

The straight line we are looking for is parallel to $y = 4x - 2$. So, its slope is $m = 4$, and its equation will be $y = 4x + n$.

As it passes by point (1, 9), this coordinates must verify $9 = 4 \cdot 1 + n$, so $n = 9 - 4 = 5$.

Our straight line is $y = 4x + 5$.

26. Write the equation of the straight line passing by point A(2,7) and whose gradient is m = -3.

27. Write the equation of the straight line passing by point A(0,0) and whose gradient is m = -7.

28. a) Calculate the slope of the straight line whose direction vector is (2,5).

b) Write the equation of the straight line passing by point A(1,2) and whose direction vector is (2,5).

29. Write the equation of the straight lines passing by point A(2,3) and B(-1,7).

30. Write the equation of the straight line passing by point A(2,-2) and whose gradient is m = -1.

31. Write the equation of the sides of the triangle whose vertices are A(2,3), B(-1,6), C(0,-2).

3. Relative positions of straight lines

Two straight lines can be: the same straight line, parallel and secant lines.

We are going to see this, remembering something you have already studied with systems of equations.

Let´s see it with an example.

Example: Represent the linear plots corresponding to both straight line equations.

a) $\begin{cases} 2x + y = 7 \\ 4x + 2y = 14 \end{cases}$ b) $\begin{cases} 2x + y = 7 \\ 2x + y = 0 \end{cases}$ c) $\begin{cases} 2x + y = 7 \\ -2x + 5y = 10 \end{cases}$

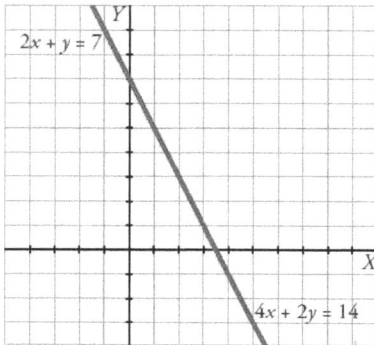

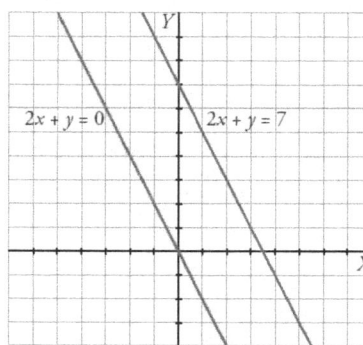

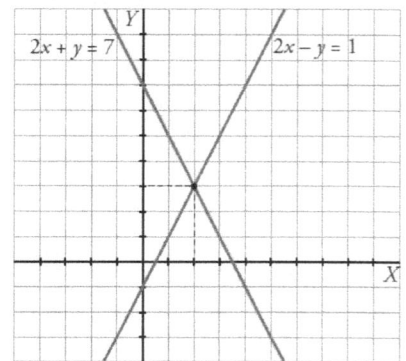

<u>Solution:</u> Feel free to give values to "x", calculate corresponding "y-values", and you must obtain plots like following:

a) $\begin{cases} 2x + y = 7 \\ 4x + 2y = 14 \end{cases}$ b) $\begin{cases} 2x + y = 7 \\ 2x + y = 0 \end{cases}$ c) $\begin{cases} 2x + y = 7 \\ -2x + 5y = 10 \end{cases}$

As you can see, both linear plots are *coincident*, so, they have infinite common points.

As you can see, both linear plots are *paralel*, so, they do not have any common point.

As you can see, both linear plots are *secant*, so, they have one common point.

QUICK IDENTIFICATION

Pay attention to the following:

a) $\begin{cases} 2x + y = 7 \\ 4x + 2y = 14 \end{cases} \rightarrow \dfrac{2}{4} = \dfrac{1}{2} = \dfrac{7}{14}$ b) $\begin{cases} 2x + y = 7 \\ 2x + y = 0 \end{cases} \rightarrow \dfrac{2}{2} = \dfrac{1}{1} \neq \dfrac{7}{0}$ c) $\begin{cases} 2x + y = 7 \\ -2x + 5y = 10 \end{cases} \rightarrow \dfrac{2}{-2} \neq \dfrac{1}{5}$

As a general rule: Given two straight lines with equations $\begin{cases} Ax + By = C \\ A'x + B'y = C' \end{cases}$		
$\dfrac{A}{A'} = \dfrac{B}{B'} = \dfrac{C}{C'}$	$\dfrac{A}{A'} = \dfrac{B}{B'} \neq \dfrac{C}{C'}$	$\dfrac{A}{A'} \neq \dfrac{B}{B'}$
Coincident	***Parallel***	***Secant***

32. Only by looking at them, indicate the relative position of the following pairs of lines:

a) $\begin{cases} 2x - y = 1 \\ 4x - 2y = 8 \end{cases}$ b) $\begin{cases} x - 2y = 5 \\ 2x - 4y = 10 \end{cases}$ c) $\begin{cases} 5x + 2y = -1 \\ 4x - y = 7 \end{cases}$ d) $\begin{cases} x - 2y = 5 \\ 2x - 4y = -3 \end{cases}$

33. Complete the following pair of straight lines so that first pair intersect at ($x = 3$, $y = -2$), second are parallel and third and fourth are coincident.

a) $\begin{cases} 3x + 2y = \ldots \\ \ldots - y = 8 \end{cases}$ b) $\begin{cases} x + y = 5 \\ 2x + 2y = \ldots \end{cases}$ c) $\begin{cases} 3x - 2y = 4 \\ 6x - 4y = \ldots \end{cases}$ d) $\begin{cases} -x + 2y = 7 \\ \ldots - 4y = \ldots \end{cases}$

34. Decide the relative position of the following pair of straight lines:

a) $\begin{cases} x - y + 3 = 0 \\ 5x + 2y - 4 = 0 \end{cases}$ b) $\begin{cases} x - y + 3 = 0 \\ 2x - 2y + 6 = 0 \end{cases}$ c) $\begin{cases} x - y + 3 = 0 \\ x - y - 2 = 0 \end{cases}$

35. Decide the relative position of the following pair of straight lines:

a) $\begin{array}{l} r \equiv x - y + 3 = 0 \\ r \equiv 2'x - 2y + 6 = 0 \end{array}$ b) $\begin{array}{l} s \equiv x - y + 3 = 0 \\ s' \equiv 5x + 2y - 4 = 0 \end{array}$ c) $\begin{array}{l} t \equiv x - y + 3 = 0 \\ t' \equiv x - y - 2 = 0 \end{array}$

4. Distances

Distance between two points

Remember the **_magnitude_** of a vector is its length. We are going to calculate it by using Pythagoras' theorem. As you can see in the figure below, coordinates v_x and v_y are the legs of a right triangle, whose hypotenuse is *magnitude* of the vector. So,

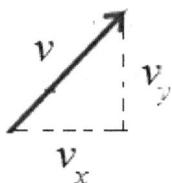

$$v^2 = v_x{}^2 + v_y{}^2 \quad \rightarrow \quad v = \sqrt{v_x{}^2 + v_y{}^2}$$

So, you can calculate the distance between two points A and B:

- **Step 1:** Calculate the coordinates of the vector **AB**: ($x_B - x_A$, $y_B - y_A$)

- **Step 2:** The distance between A and B is the magnitude of vector **AB**:

$$d(A, B) = \sqrt{(x_B - x_A)^2 + (y_B - y_A)^2}$$

Example: Calculate the distance between the points C(2, 5) and D(0, 4).

Solution:

a) **CD** = $(0 - 2, 4 - 5) = \boxed{(-2, -1)}$

b) $d = \sqrt{(-2)^2 + (-1)^2} = \sqrt{4+1} = \boxed{\sqrt{5}}$

Distance between a point and a straight line

Given a point $A(x_0, y_0)$ and the straight line with equation $Ax + By + C = 0$, the **perpendicular distance** between them is:

$$d(A,r) = \frac{A \cdot x_0 + B \cdot y_0 + C}{\sqrt{A^2 + B^2}}$$

Example: Calculate the distance between the point P(0,6) and the straight line $\dfrac{x-2}{3} = \dfrac{y+1}{5}$

Solution: First of all, you must rewrite the equation:

$$\frac{x-2}{3} = \frac{y+1}{5} \rightarrow 5 \cdot (x-2) = 3 \cdot (y+1) \rightarrow 5x - 10 = 3y + 3 \rightarrow 5x - 3y - 13 = 0$$

Now, you can use the previous expression: $d(A,r) = \dfrac{5 \cdot 0 - 3 \cdot 6 - 13}{\sqrt{5^2 + (-3)^2}} = \dfrac{-31}{\sqrt{34}}$

As you know, a distance cannot be a negative number. So, we must calculate its absolute value, as well as rationalize it:

$$d(A,r) = \left| \frac{-31}{\sqrt{34}} \right| = \boxed{\frac{31\sqrt{34}}{34}}$$

Exercises

36. Calculate the distance between P and Q:

    a) P(3, 5), Q(3, −7)   b) P(−8, 3), Q(−6, 1)

    c) P(0, −3), Q(−5, 1)   d) P(−3, 0), Q(15, 0)

37. a) Find out the mid-point segment-line with extremes A(−2, 0) and B(6, 4).
    b) Check that the distances from the mid-point to both extremes are equal.

38. Check that the triangle with vertices A(−1, 0), B(3, 2), C(7, 4) is an isosceles one. Which are the equal sides?

39. By using the Pythagoras´ Theorem, check that the triangle with vertices A(−2, −1), B(3, 1) and C(1, 6) is a right one.

Review exercises

1. Work out the equations of the straight line passing by the point A(-3,-2) and whose direction vector is $\vec{u}$ (1,4).

2. Work out the equations of the straight line passing by the point A(-3,6) and whose direction vector is $\vec{u}$ (3,-1).

3. Work out the equations of the straight line passing by the point A(2,8) and whose direction vector is $\vec{u}$ (-1,9).

4. Write the equation of the straight line parallel to r: 3x+2y-4=0 that passes by the point A(2,3).

5. Given the straight lines: 2x-y+6=0   and   x+2y-3=0, are they perpendicular?

6. Write the equation of the straight line perpendicular to r: 2x+y-6=0 that passes by the point A(4,5).

7. Write the equation of the following straight lines:

   a) Passes by P(1,-2) and it is parallel to the line 2x-3y+8=0

   b) Passes by Q(3,1) and it is perpendicular to the line x+3y-5=0

8. Decide the relative position of the following pair of straight lines:

   a) 3x+2y-2=0 and 6x+4y-4=0   b) 6x-2y=0 and 6x-2y=15   c) 3x-y+4=0 and   2x+3y-2=0.

9. Write the equation of the straight line that passes by the point (-2,5) and it is:

   a) Parallel to X-axis.   b) Parallel to Y-axis   c) Passes by (0,0).

10. Given the straight lines r: 3x+y-11=0 and s: x+2y-7=0, write the equation of the straight line that passes by the intersection point of the straight lines s r and s and by the point A(-1,2).

11. Write the equation of the straight line that passes by the intersection point of the straight lines:

   3x+4y-10=0   4x-3y-5=0   and it passes by the point P(-3,2)

12. Calculate *a* and *b* so that straight lines:

   3x+by-8=0   ax-3y+12=0   intersect at the point P(2,-3)

13. Write the equation of the straight line that passes by the point P(5,2) and has the same slope than the straight line 3x+4y-5=0

14. Find out the coordinates of the vertices of the triangle whose sides are at the straight lines

   x-y-1=0   x+y+2=0   y=3x+2

15. Write the equation of the diagonals and their intersection point of the quadrilateral having its sides at the straight lines *r,s,t,u,* where:

   r: 3x+4y-8=0   t: x-2y+12=0   s: 2x+y+5=0   u: 2x+y+2=0

16. Straight lines mx+2y=3 and 5x+ny=7 intersect at the point (-1,3). Calculate m and n.

17. Write the equation of the straight line that passes by the point (0,0) and it is parallel to the straight line that passes by the points A(1,2) and B(3,-4).

18. Work out the mediatrix of the segment line whose extremes are:

a) A(2,3) and B(5,4)   b) M(-1,2) and N(0,5)   c) P(-1,0) and Q(6,3).

19. Determine the equation of the straight line that passes by the intersection of the lines with equations 4x+6y-5=0 and x-2y-3=0, and it is parallel to the line with equation 4x-5y-12=0.

20. Calculate the distance between the following points:

a) A(2,-1) and B(3,4)   b) M(6,-2) y N(3,8)   c) P(-8,0) y Q(-3,9)

21. Given the straight line with equation r: x-4y+7=0, find out:

a) Distance from P(2,5) to r   b) Distance from origin to r.

22. Calculate the distance between the straight lines r:2x+3y-5=0 and r: 2x+3y-7=0

23. Calculate the distance between the straight lines r and r′ whose equations are:

$$r:\begin{cases} x = 2-3t \\ y = 1+t \end{cases} \qquad r':\frac{x+3}{-3} = \frac{y+5}{1}$$

24. Calculate the distance between the following points and straight lines:

a) P(2,-1) r: x+y-4=0   b) Q(1,3) $r:\dfrac{x-2}{-1} = \dfrac{y+3}{2}$

c) R(4,0) $r:\begin{cases} x = 2-3t \\ y = t \end{cases}$

25. Calculate the distance between the following straight lines:

a) x+6y-5=0   2x-y+3=0   b) 2x-y+6=0   2x-y+15=0

c) x-2y+4=0 $\dfrac{x-1}{2} = \dfrac{y+5}{1}$ d) $\dfrac{x-2}{5} = \dfrac{y-1}{3}$ $\dfrac{x}{-1} = \dfrac{y+6}{2}$

Unit 8.- Statistics

1. Introduction to statistics. First concepts

Statistics have become an important part of everyday life. We are confronted by them in newspapers and magazines, on television and in general conversations. We encounter them when we discuss the cost of

living, unemployment, medical breakthroughs, weather predictions, sports, politics and the state lottery. Although we are not always aware of it, each of us is an informal statistician. We are constantly gathering, organizing and analysing information and using this data to make judgments and decisions that will affect our actions. In this unit we will try to improve the students understanding of the elementary

topics included in statistics. The unit will begin by discussing terms that are commonly used in statistics. It will then proceed to explain and construct frequency distributions, histograms, frequency polygons and cumulative frequency polygons. Next, the unit will define and compute measures of central tendency including the mean, median and mode of a set of numbers. Measures of dispersion including range and standard deviation will be discussed.

Population and sample

Statistics is a branch of mathematics in which groups of measurements or observations are studied. Statistics deals with methods used to collect organize and analyse numerical facts. Its primary concern is to describe information gathered through observation in an understandable and usable manner.

Throughout the study of statistics certain basic terms occur frequently. Some of the more commonly used terms are defined below.

Statistics work begins when collecting data. As you can imagine, if you want to study heights of citizens in a town, you cannot measure all the *population*. You must choose an appropriate *sample*.

A *population* is a complete set of items that is being studied. It includes all members of the set. The set may refer to people, objects or measurements that have a common characteristic.

A *sample* is a relatively small group of items selected from a population. If every member of the population has an equal chance of being selected for the sample, it is called a *random sample*.

For example, the owner of a screw factory wants to make a quality control. He picks up 1 out of every 100 produced screws and then he analyses them.
- The *population* is the total number of the screws of the factory.
- The *sample* is 1% of the population.
- The *individuals* are each one of the screws.

2. Statistical variables

Data are numbers or measurements that are collected. Data may include numbers of individuals that make up the census of a city, ages of pupils in a certain class, temperatures in a town during a given period of time, sales made by a company, or test scores made by ninth graders on a standardized test.

Variables are characteristics or attributes that enable us to distinguish one individual from another. They take on different values when different individuals are observed.

Some variables are height, weight, age and price. Variables are the opposite of *constants,* whose values never change.

Statistical variables can be classified in several ways: quantitative and qualitative variables.

Quantitative and qualitative variables

- A **quantitative or numerical variable** is that one whose data are numbers. For example, heights, number of books, incomes in €, etc.

- A **qualitative or categorical variable** is that one whose data have labels (i.e. words). For example, your favourite musical group, a list of the products bought by different families at a grocery store, eyes colour, etc.

Discrete and continuous variables

Numerical variables can be classified into discrete and continuous variables.

- **Discrete variables** are those whose data are whole numbers, and are usually a count of objects. For instance, number of pets in a house, number of children in a family, ….. It does not make sense to have 3.5 children.

- **Continuous variables** are those that may take any real value. For example, the amount of time a group of children spent watching TV, heights, weights, etc. Your weight may be 74 kg, 74.5 kg or 74.568 kg.

<table>
<tr>
<td rowspan="2">Exercises</td>
<td>

1. For each of the following cases, indicate what are the population, the variable and the type of variable.

    a) Weight of babies that were born last year in Dublin.

    b) Favourite subject for the students in school.

    c) Number of pets in French households.

    d) Political party that the Spanish electors are going to vote for in the next local elections.

    e) Weekly time that students from 12 to 16 spend on reading in Italy.

</td>
</tr>
</table>

3. Frequency distributions. Frequency tables

Large groups of data have little value until they have been placed in some kind of order. Usually, measurements are listed in ascending or descending order. Such a group is a *distribution*. A **frequency distribution** is a table in which measurements and the *frequency* or total number of times that each item occurs is recorded.

Example : The number of televisions in each house of my street is shown in the frequency table:

Number of TVs	Number of houses
0	1
1	5
2	12
3	9
4	1

a) Calculate the number of houses in my street.
b) Calculate the total of number of televisions in my street.

Solution:

a) The total number of houses is: $1 + 5 + 12 + 9 + 1 =$
= 28 houses.

b) The total number of televisions is: $(0 \cdot 1) + (1 \cdot 5) + (2 \cdot 12) + (3 \cdot 9) + (4 \cdot 1) = 0 + 5 + 24 + 27 + 4 =$
= 60 televisions.

Absolute and relative frequencies

The **absolute frequency** is the number of times a determined value of a variable occurs.

We will write f_i the absolute frequency of x_i.

We have to know that $f_1 + f_2 + ... + f_n = N$, where N is the total number of data. (n is the number of different data).

The **relative frequency** is the absolute frequency divided by the total number of observations or total number of data.

We will write h_i the relative frequency of x_i.

And we have $h_1 + h_2 + ... + h_n = 1$.

Exercises

2. In the list below is shown the qualification in Maths of 20 students in a class:

$$3 \quad 3 \quad 3 \quad 4 \quad 5$$
$$5 \quad 5 \quad 5 \quad 5 \quad 6$$
$$6 \quad 7 \quad 7 \quad 7 \quad 7$$
$$8 \quad 9 \quad 9 \quad 9 \quad 9$$

a) Build a table writing the absolute and relative frequencies of this set of data.

x_i	Absolute frequency f_i	Relative frequency h_i
...		
...		
...		

b) Calculate the sum of all the absolute frequencies and of all relative frequencies.

3. Mrs. Parker asked her students about their favourite subject in the school, and the answers were:

Maths	English	Sciences	English	Maths
History	English	Music	English	Maths
English	Maths	English	History	English
English	English	English	English	Maths
History	Sciences	Maths	Maths	Sciences

Make a table and show the absolute and relative frequencies of these data.

4. In thirty shots, a man makes the following scores:

5 2 2 3 4   4 3 2 0 3   0 3 2 1 5   1 3 1 5 5   2 4 0 0 4   5 4 4 5 5

Summarize the absolute and relative frequencies of the different scores.

5. There are 44 students in a group. Each student plays either hockey or tennis but not both. Complete the table.

	Hockey	Tennis	Total
Girls	8		20
Boys	18		24

Cumulative absolute and relative frequency

- **Cumulative Absolute Frequency** of a datum x_i is the sum of the absolute frequencies of values less or equal than it. We write F_i.

$$F_i = f_1 + f_2 + ... + f_i$$

- **Cumulative Relative Frequency** of a datum x_i is the sum of the relative frequencies of values less or equal than it. We write H_i. It is the quotient between the cumulative absolute frequency and the total number of data.

$$H_i = h_1 + h_2 + ... + h_i$$

6. It is shown the number of brothers and sisters of the 25 students:

0 2 1 0 1   1 1 2 1 3   0 1 1 1 1   0 1 1 2 3   2 1 0 4 2

Complete the table with the different frequencies:

x_i	f_i	F_i	h_i	H_i
...				
...				
...				

7. We asked 18 students of 2nd E.S.O about the age of their parents. The answers were the following: 40 42 44 47 44   41 43 46 42 46   44 45 47 42 45   40 43 47

Complete a frequency table with the different frequencies.

4. Statistical graphs

Many people find it easier to obtain information from pictures than from written material. Statisticians display mathematical relationships with diagrams and graphs. From these pictures numerical data can be summarized clearly and easily.

Bar diagrams

When the data of a frequency distribution have not been grouped in intervals, they can be represented on a *bar diagram*. A bar diagram illustrates the pattern of a distribution. It clearly shows whether the data are spread out evenly or if they tend to cluster about a point.

To build a bar diagram, list the measurements, from lower to higher, horizontally across the bottom of the graph. On the left side vertically list the frequencies or number of times that the measurements occur. Finally, on each measurement, draw a bar with height equal to its frequency. We draw all the bars with the same width.

Bar diagram is suitable for discrete variable (numerical or categorical).

For example, following figure shows a bar diagram indicating frequencies of vowels in a sentence in English:

Vowels in a sentence in English

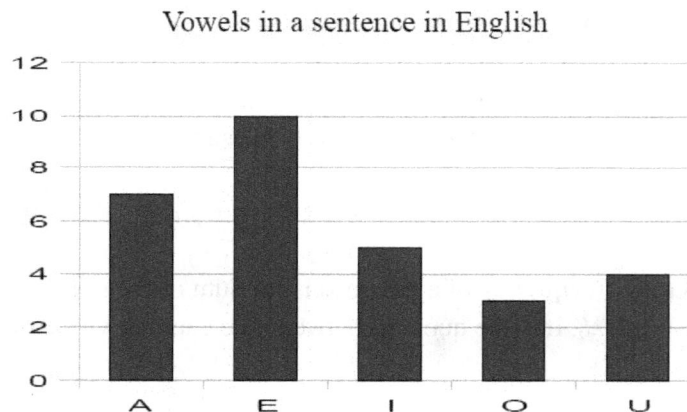

Histograms

The **histogram** is used for variables whose values are numerical and measured on an interval scale (usually for continuous variables).

A histogram divides up the range of possible values in data set into classes or groups. For each group, a rectangle is constructed with an area proportional to the frequency, (if the bars have identical widths, the height of each bar corresponds to the frequency, but if they do not, if each bar has a width a_i, then the heights will be $\dfrac{f_i}{a_i}$, so that its area is f_i).

Example: Mario decided to collect data about the height of his classmates in the school. These are the data of 40 children (in cm):

163	167	165	159	164	168	161	164	163	164
165	163	167	165	164	164	168	161	164	165
163	164	170	160	157	167	165	172	165	167
164	164	168	151	164	163	164	155	158	162

<u>Solution:</u>

It is useful to make a group frequency table for this case:

Interval	Tally	Frequency
[155-160)	////	4
[160-165)	### ### ### ### /	21
[165-170)	### ### ///	13
[170-175)	//	2

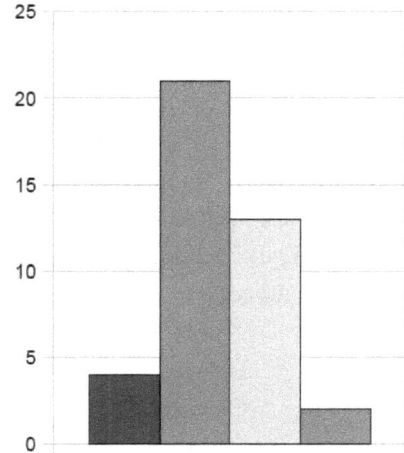

---

8. These are the heights, in cm, of 40 people.

153	134	155	142	140	163	150	135	170	156	171	161	141
153	144	163	140	160	172	157	136	160	134	154	176	154
173	179	160	152	170	148	151	165	138	143	147	144	156
139

Complete the following frequency table and draw the histogram.

Height (cm)	Number of people
[130,140)	
[140,150)	
[150,160)	
[160,170)	
[170,180)	

9. This frequency table shows the times for 50 runners in a Marathon.

Time (hours)	Number of runners
[1,2)	0
[2,3)	15
[3,4)	23
[4,5)	18
[5,6)	8

Draw a histogram to show the times.

Exercises

Pie charts

A **pie chart** is a way of summarizing a set of categorical data. It is a circle which is divided into sectors. Each sector represents a particular category. The area of each sector is proportional to the number of cases in that category.

Example: We ask 240 people to name their favourite fruit. With these results, draw a pie chart to illustrate the information.

Fruit	Apple	Banana	Orange	Other
Number of people	50	80	72	38

Solution: First, we calculate the angle for one person: $360° : 240 = 1.5°$

Then, we calculate the angles of each category:

Apple: $50 \cdot 1.5° = 75°$

Banana: $80 \cdot 1.5° = 120°$

Orange: $72 \cdot 1.5° = 108°$

Other: $38 \cdot 1.5° = 57°$

And then, we measure, colour and label the sectors:

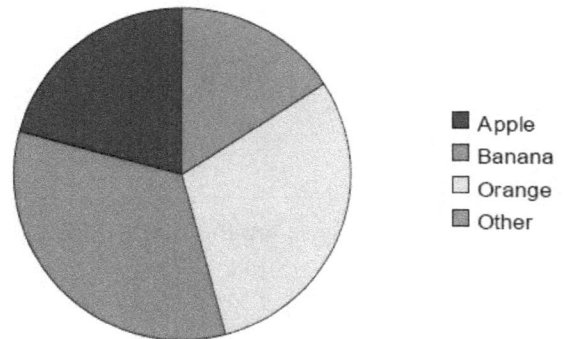

Favourite Fruit

- Apple
- Banana
- Orange
- Other

Exercises

10. The Manchester United football team plays 36 matches in a season. They win 15 matches, and they draw just 8 matches.

a) Calculate the number of matches that they lose.

b) Calculate the angle one match represents in a pie chart.

c) Calculate the angle of each category in the pie chart.

d) Draw a pie chart to show the information.

11. The weather record for 60 days is shown in the frequency table.

Weather	Number of days
Sunny	15
Cloudy	18
Rainy	14
Snowy	3
Windy	10

a) Calculate the angle one day represents in a pie chart.

b) Calculate the angle of each category in the pie chart.

c) Draw a pie chart to show the data.

Frequency polygons

A **frequency polygon** is a line graph which can be used to represent the frequency of a set of numbers. It is formed by connecting a series of points. The abscise of each point is the midpoint of the interval in which the point lies. The ordinate of each point is the frequency for the interval. The polygon is closed at each end by drawing a line from the endpoints to the horizontal axis at the midpoint of the next interval.

Example: In an intelligence test, 200 people have been examined. Results are given in the following table.

Interval	Mark
30 – 40	6
40 – 50	18
50 – 60	76
60 – 70	70
70 – 80	22
80 – 90	8

Draw a histogram to represent them and build, also, the frequency polygon.

Solution:

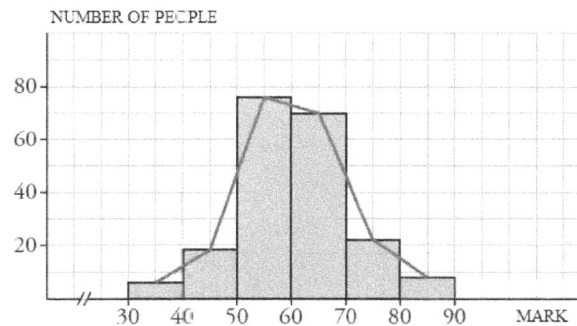

12. These are the heights, in cm, of 40 people.

153	134	155	142	140	163	150	135	170	156	171	161	141
153	144	163	140	160	172	157	136	160	134	154	176	154
173	179	160	152	170	148	151	165	138	143	147	144	156
139												

Complete the following frequency table and draw the frequency polygon:

Height (cm)	Number of people
[130,140)	
[140,150)	
[150,160)	
[160,170)	
[170,180)	

Exercises

5. Measures of central tendency and dispersion

In the following sections, *xi* will denote an isolated value as well as the mi-point of the interval classfor grouped data.

Mean

The *average* or *mean* of a list of numbers is the total of all values divided by the number of values.

To calculate the mean, we can use the absolute frequencies of values, multiplying every value by its absolute frequency, then adding all these products and finally dividing by the number of values. It is shown in the formula:

$$\bar{x} = \frac{x_1 \cdot f_1 + x_2 \cdot f_2 + \ldots + x_n \cdot f_n}{N} = \frac{\sum x_i \cdot f_i}{N}$$

Example: Find the mean of 10, 11, 7 and 8:   Solution: $\bar{x} = \frac{10 + 11 + 7 + 8}{4} = \frac{36}{4} = \boxed{9}$.

Example: Find the mean of the following values:

$$4\ 4\ 4\ 4\ 5 \quad 5\ 6\ 6\ 6\ 6 \quad 6\ 7\ 8\ 8\ 8 \quad 9\ 9\ 9\ 9\ 10$$

Solution: We can use a frequency table:

x_i	f_i
4	4
5	2
6	5
7	1
8	3
9	4
10	1

$$\bar{x} = \frac{4 \cdot 4 + 5 \cdot 2 + 6 \cdot 5 + 7 \cdot 1 + 8 \cdot 3 + 9 \cdot 4 + 10 \cdot 1}{20} = \frac{16 + 10 + 30 + 7 + 24 + 36 + 10}{20} = \frac{133}{20} = \boxed{6.65}.$$

Exercises

13. Calculate the mean of: 5, 3, 54, 93, 83, 22, 17 and 19.

14. Calculate the mean of values given as a table:

x_i	f_i
2	5
3	5
4	7
5	8

Median, quartiles and percentiles

- The median, *Me,* of a list of values is found by ordering them from lower to higher. Once they are arranged, median is the central value. Median is the value that has as 50% of the values lower than it.
- Quartiles, *Q₁* and *Q₃* are the values that have 25% and 75% of the values lower than them.
- A percentile, *Pi,* is the value that has a *i%* of the values lower than it.

Example: Money weekly given by parents to a group of students is:

$$10 - 15 - 12 - 20 - 25 - 18 - 12 - 30 - 22 - 19 - 18 - 15 - 13 - 20 - 24$$

Calculate the median, the quartiles and the percentile 40.

Solution: First of all, we have to arrange the data:

$$10 - 12 - 12 - 13 - 15 - 15 - 18 - 18 - 19 - 20 - 20 - 22 - 24 - 25 - 30$$

There are 15 students:

- Me: $15 \cdot \frac{1}{2} = 7.5 \rightarrow$ *Me* will be between 7th and 8th data. As both of them are 18, $\boxed{Me = 18}$.

- Q_1: $15 \cdot \frac{1}{4} = 3.75 \rightarrow$ *Q₁* will be between 3rd and 4th data. As they are different, $Q_1 = \frac{12 + 13}{2} = \boxed{12.5}$.

- Q_3: $15 \cdot \frac{3}{4} = 11.25 \rightarrow$ *Q₃* will be between 11th and 12th data. As they are different, $Q_3 = \frac{20 + 22}{2} = \boxed{21}$.

- P_{40}: $15 \cdot \frac{40}{100} = 6 \rightarrow$ *P₄₀* will be the 6th value, $\boxed{P_{40} = 15}$.

Mode

The **mode** in a list of numbers is the most repeated number (or numbers). It is the value with the highest absolute frequency.

Example: Find the mode of following sets: a) 2, 3, 3, 6, 4, 3, 2, 5, 6, 3;   b) 2, 2, 2, 3, 4, 4, 4, 5, 6, 6.

Solution:

   a) The mode is 3, because it is the number that appears most often in the list. $\boxed{Mo = 3}$.
   b) We have two modes in this case, 2 and 4, because they appear three times. Both numbers have the highest absolute frequency. $\boxed{Mo = 2, 4}$.

15. Find the mean, median and mode for the following data:

10, 12, 13, 12, 13, 10, 14 and 13.

16. Twenty families are asked about how many children they have. These are the answers:

3 3 4 1 2 3 2 5 1 0

2 2 3 2 4 2 5 3 4 3

a) Complete the table with the frequencies: f_i, h_i, F_i and H_i.

b) Find the mean, median and mode.

17. Eight people work in an office. They are paid hourly rates of

$12, $15, $15, $14, $13, $14, $13, $13

Find: a) the mean, b) the median, c) the mode.

18. The following table shows the number of children that 100 families in a town have.

Children	1	2	3	4	5	6
Families	48	25	16	4	5	2

Find the mean, mode, median, Q_1, Q_3 and P_{30}.

19. A gardener buys 10 packets of seeds from two different companies. Each pack contains 20 seeds and he records the number of plants which grow from each pack.

Company A: 20 5 20 20 20 6 20 20 20 8

Company B: 17 18 15 16 18 18 17 15 17 18

a) Find the mean, median and mode for each company's seeds.

b) Which company does the mode suggest is best?

c) Which company does the mean suggest is best?

20. In a beauty contest, the scores awarded by eight judges were:

5.9   6.7   6.8   6.5   6.7   8.2   6.1   6.3

Determine: a) the mean, b) the median, c) the mode.

d) Only six scores are to be used. Which two scores may be omitted to leave the value of the median the same?

21. The students in a class state how many children there are in their family. The numbers they state are given below.

1, 2, 1, 3, 2, 1, 2, 4, 2, 2, 1, 3, 1, 2,

2, 2, 1, 1, 7, 3, 1, 2, 1, 2, 2, 1, 2, 3

a) Find the mean, median and mode for this data.

b) Which is the most sensible average to use in this case?

22. A manager keeps a record of the number of calls she makes each day on her cell phone.

Number of calls per day	0	1	2	3	4	5	6	7	8
Frequency	3	4	7	8	12	10	14	3	1

Calculate the mean number of calls per day.

23. A class conduct an experiment in biology. They place a number of 1 m by 1 m square grids on the playing field and count the number of plants in each grid. The results obtained are given below.

6	3	2	1	3	2	1	3	0	1
0	3	2	1	1	4	0	1	2	0
1	1	2	2	2	4	3	1	1	1
2	3	3	1	2	2	2	1	7	1

a) Calculate the mean number of plants.

b) How many times was the number of plants seen greater than the mean?

24. Hannah drew this bar chart to show the number of repeated cards she got when she opened packets of football stickers. Calculate the mean number of repeats per packet.

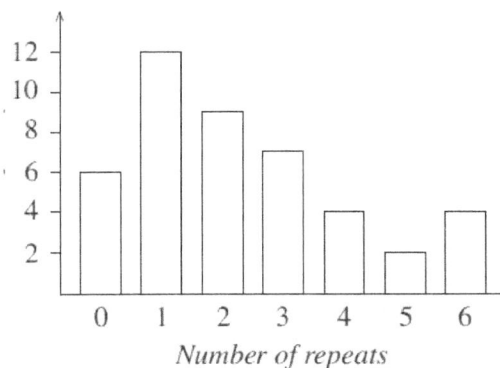

Number of repeats

25. The number of days that students were missing from school due to sickness in one year was recorded. Estimate the mean.

Number of days off sick	(1 − 5)	(6 − 10)	(11 − 15)	(16 − 20)	(21 − 25)
Frequency	12	11	10	4	3

26. The table shows the distribution of scores or 40 students on a Mathematics test. Estimate the mean score obtained on the test.

Score	10 - 12	13 - 15	16 - 18	19 - 21	22 - 24
Frequency	4	6	13	9	8

Estimate the mean score obtained on the test.

165

Median, quartiles and percentiles for grouped data

Example: In a petrol station, they are recording the number of cars depending on the time. This is the result:

Hours:	[0, 4)	[4, 8)	[8, 12)	[12, 16)	[16, 20)	[20, 24)
Number of cars:	6	14	110	120	150	25

Calculate the median and Q_3.

Solution: Build the cumulative frequency table:

Interval higher extreme	Fi	Hi	Hi (%)
0	0	0	0
4	6	0,0141	1,41
8	20	0,0471	4,71
12	130	0,3059	30,59
16	250	0,5882	58,82
20	400	0,9412	94,12
24	425	1	100

Now, we are working on the frequency polygon, building the accurate triangles, and applying what we¡e already know about similar triangles:

Me:

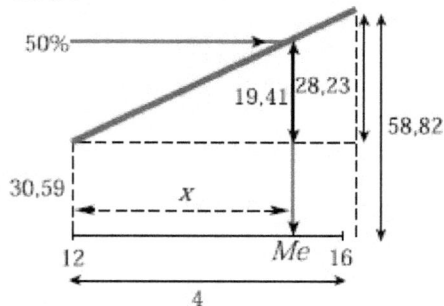

Q₃:

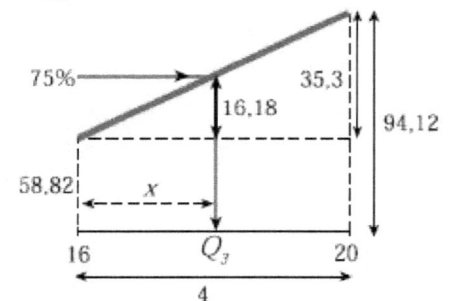

$$\frac{28,23}{4} = \frac{19,41}{x}$$
$$x = 2,75$$
$$Me = 12 + 2,75 = \boxed{14,75}$$

$$\frac{35,3}{4} = \frac{16,18}{x}$$
$$x = 1,83$$
$$Q_3 = 16 + 1,83 = \boxed{17,83}$$

27. Time taken by employees of a firm to go from home to the office is shown in the next table:

Time (min)	[0, 15)	[15, 30)	[30, 45)	[45, 60)	[60, 75)	[75, 90)
Workers	10	23	32	5	6	4

Calculate Me, Q_1 and Q_3.

28. The table below gives data on the heights, in cm, of 51 children.

Class Interval	$140 \leq h < 150$	$150 \leq h < 160$	$160 \leq h < 170$	$170 \leq h < 180$
Frequency	6	16	21	8

Estimate the mean height and the median.

29. Incomes, in M€, of 500 firms, are shown in the following table:

Incomes	[1, 2)	[2, 3)	[3, 4)	[4, 5)	[5, 6)	[6, 7)
Firms	50	80	170	90	56	54

Calculate the mean, Me, Q_1 and Q_3.

30. The weights of a number of students were recorded in kg.

Weight	$30 \leq w < 35$	$35 \leq w < 40$	$40 \leq w < 45$	$45 \leq w < 50$	$50 \leq w < 55$
Frequency	10	11	15	7	4

Estimate the mean weight and the median.

31. We have recorded the last digit of the phone number of a quantity of people.

Last digit	0	1	2	3	4	5	6	7	8	9
Times	28	35	28	29	45	32	37	45	25	61

Calculate Me, Q_1, Q_3 and P_{90}.

Range

Range is defined as the difference between the largest and smallest sample values.

Range depends only on extreme values and provides no information about how the remaining data is distributed.

Mean deviation

We can easily calculate how much each number deviates from the mean.

Mean deviation is the mean of these deviations. $MD = \dfrac{|x_1 - \bar{x}| + |x_2 - \bar{x}| + \dots\dots + |x_n - \bar{x}|}{n}$

Variance

The Variance (σ^2) is a measure of how spread our numbers are. Its symbol is the squared of Greek letter sigma.

The **variance** is calculated as the average of the **square** of the differences from the mean. Its formula is:

$$\sigma^2 = \frac{(x_1 - \bar{x})^2 + (x_2 - \bar{x})^2 + \dots + (x_n - \bar{x})^2}{n} = \frac{\sum(x_i - \bar{x})^2}{n}$$

There is an equivalent formula to calculate variance. Sometimes it is easier to use:

$$\sigma^2 = \frac{x_1^2 + x_2^2 + \dots + x_n^2}{n} - (\bar{x})^2 = \frac{\sum x_i^2}{n} - (\bar{x})^2$$

Standard deviation

The **standard deviation** is defined as the square root of variance. So, it is denoted by σ.

$$\sigma = \sqrt{\sigma^2}$$

Example: Calculate the dispersion measures of this weights distribution:

$$83, 65, 75, 72, 70, 80, 75, 90, 68, 72$$

Solution: First of all, we arrange data: 65, 68, 70, 72, 72, 75, 75, 80, 83, 90.

And, calculate the mean, because we will need it after:

$$\bar{x} = \frac{65 + 68 + 70 + 72 + 72 + 75 + 75 + 80 + 83 + 90}{10} = \frac{750}{10} = 75$$

- Range: $90 - 65 = \boxed{25}$.

- Mean deviation: $MD = \dfrac{|68-75|+|68-75|+|70-75|+|72-75|+\ldots\ldots\ldots+|90-75|}{10} = \dfrac{56}{10} = \boxed{5.6}$.

- Variance: $MD = \dfrac{(65-75)^2+(63-75)^2+(70-75)^2+\ldots\ldots+(90-75)^2}{10} = \dfrac{506}{10} = \boxed{50.6}$.

- Standard deviation: $\sigma = \sqrt{\sigma^2} = \sqrt{50.6} = \boxed{7.11}$.

Variation coefficient

Standard deviation gives us information about dispersion of data. But, sometimes, we want to compare dispersions of different sets of data. If magnitudes of data in different sets of data are very distinct, standard deviation might not be useful.

For example, imagine we have two sets of weights: set A are weights of cars and set B are weights of oranges. Obviously, standard deviation of set B is going to be less than in set A, but that does not mean that in this set there is a lower dispersion of data, it is just because its data a quite lower.

In these cases, we need a new parameter, the *variation coefficient*.

The **variation coefficient** is defined as $\qquad VC = \dfrac{\sigma}{x}$

As you can imagine, the higher VC is, the higher dispersion is.

Example: In three restaurants, A, B and C, we are investigating salaries (€ per weekend).

	A	B	C
Mean	943	132	37
Standard deviation	148	22	12

In what restaurant do they have the most homogenous salaries?

Solution:

$$VC_A = \dfrac{943}{148} = 6.37 \qquad\qquad VC_B = \dfrac{132}{22} = 6 \qquad\qquad VC_C = \dfrac{37}{12} = 3.08$$

So, as the lowest VC has been obtained for restaurant C, this is the one having the most homogenous salaries.

32. Find the mean, range, variance, standard deviation and coefficient of variation of the numbers

$$6, 7, 8, 5, 9$$

33. The table below gives the number of road traffic accidents per day in a small town.

Accidents per day	0	1	2	3	4	5	6
Frequency		5	8	6	3	2	1	1

Find the mean and standard deviation of this data.

34. Find the mean and standard deviation of each set of data given below.

A 51 56 51 49 53 62

B 71 76 71 69 73 82

C 102 112 102 98 106 124

35. Calculate the coefficient of variation of the previous sets.

36. Two machines, A and B, fill empty packets with soap powder. A sample of packets was taken from each machine and the weight of powder (in kg) was recorded.

A 2.27 2.31 2.18 2.2 2.26 2.24

B 2.78 2.62 2.61 2.51 2.59 2.67 2.62 2.68 2.70

a) Find the mean and standard deviation for each machine.

b) Which machine is most consistent?

37. Two groups of students were trying to find the acceleration due to gravity. Each group conducted 5 experiments.

Group A	9.4	9.6	10.2	10.8	10.1
Group B	9.5	9.7	9.6	9.4	9.8

Find the mean and standard deviation for each group, and comment on their results.

38. In two groups, 3rd A and 3rd B, we have done the same exam. Results are in the following table.

	Mean	Std. Dev.
3rd **A**	5.8	2.9
3rd **B**	6.3	1.2

a) Calculate their variation coefficients and indicate in which group results have more dispersion.

b) In a group, there were 6 nines and 5 ones, while in the other one, there were only 1 nine and 2 ones. Could you indicate these groups?

6. Two-dimensional distributions

So far we have studied a single characteristic of a population *(size, weight, ...)*, but we could study several of them simultaneously. For example:

    1. Weight and height of a sample of 100 people.

    2. Number of hours students spend watching television and academic results.

    4. Production and sales of a factory.

This type of statistical variable is called ***two-dimensional statistics***.

Often, we want to study the relationship between two characteristics of a population. This is the object of the ***"linear regression"***.

Two-dimensional variables

They are those resulting from the observation of a phenomenon on two ***modalities***, the pair ***(X, Y)***, where X is a one-dimensional variable taking values x_1, x_2, ... x_n, and Y is another one-dimensional variable which takes the values y_1, y_2, ... y_n. So, the two-dimensional statistical variable (X, Y) takes values: (x_1, y_1), (x_2, y_2), ...(x_n, y_n).

If we represent these pairs in a system of Cartesian axes a set of points on the plane called ***scatter plot*** or ***point cloud*** is obtained.

Two-dimensional frequency tables

Example: We have measured ages and weighs of 5 children and these are the results:

Edad (años)	2	4'5	6	7'2	8
Peso (Kg)	15	19	25	33	34

Its scatter plot is:

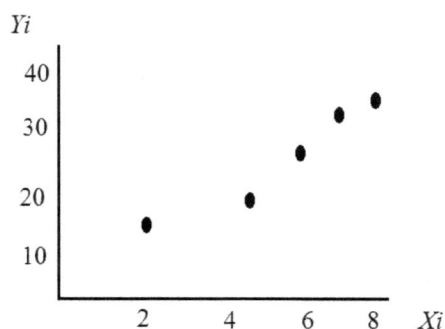

Example: 50 families have been classified according to the sex of their children. Let X the number of boys and Y the number of girls. The results are:

y\x	0	1	2	3	4	5	6	
0	2	-	4	3	1	-	-	10
1	3	-	9	-	-	3	-	15
2	-	6	-	6	-	-	1	13
3	1	4	-	-	2	1	-	8
4	-	-	2	-	1	-	-	3
5	-	-	-	1	-	-	-	1
	6	10	15	10	4	4	1	50

This kind of table are called *2-input data table*.

7. Calculation of statistical parameters

2-dimensional data are organized in tables as following one:

VARIABLE X x_i	VARIABLE Y y_i	ABSOLUTE FREQUENCY f_i
x_1	y_1	f_1
x_2	y_2	f_2
x_3	y_3	f_3
....	....	....
x_n	y_n	f_n
		$N = \Sigma f_i$ = Total quantity of data

Expressions to calculate mean, variance and standard deviation are:

	Variable X	Variable Y
Mean	$\bar{x} = \dfrac{\sum\limits_{i=1}^{n} x_i \cdot f_i}{N}$	$\bar{y} = \dfrac{\sum\limits_{i=1}^{n} y_i \cdot f_i}{N}$
Variance	$\sigma_x^2 = S_x^2 = \dfrac{\sum\limits_{i=1}^{n} f_i \cdot (x_i - \bar{x})^2}{N}$	$\sigma_y^2 = S_y^2 = \dfrac{\sum\limits_{i=1}^{n} f_i \cdot (y_i - \bar{y})^2}{N}$
	$\sigma_x^2 = S_x^2 = \dfrac{\sum\limits_{i=1}^{n} f_i \cdot x_i^2}{N} - \bar{x}^2$	$\sigma_y^2 = S_y^2 = \dfrac{\sum\limits_{i=1}^{n} f_i \cdot y_i^2}{N} - \bar{y}^2$

Remember standard deviation, σ, equals squared root of variance, $\sigma = \sqrt{\sigma^2}$

In this case, we have $\sigma_x = \sqrt{\sigma_x^2}$ and $\sigma_y = \sqrt{\sigma_y^2}$.

Covariance

Covariance is calculated as: $\sigma_{xy} = S_{xy} = \dfrac{\sum\limits_{i=1}^{n} f_i \cdot (x_i - \bar{x})(y_i - \bar{y})}{N} = \dfrac{\sum\limits_{i=1}^{n} f_i \cdot x_i y_i}{N} - \bar{x} \cdot \bar{y}$

Example: We have measured ages and weighs of 5 children and these are the results:

Age (years)	2	4'5	6	7'2	8
Weigh (kg)	15	19	25	33	34

Calculate means, variances, standard deviations and covariance.

Solution: First, build the following table:

x_i	y_i	f_i	$x_i \cdot f_i$	$y_i \cdot f_i$	$x_i^2 \cdot f_i$	$y_i^2 \cdot f_i$	$x_i \cdot y_i \cdot f_i$
2	15	1	2	15	4	225	30
4'5	19	1	4'5	19	20'25	361	85'5
6	25	1	6	25	36	625	150
7'2	33	1	7'2	33	51'84	1089	237'6
8	34	1	8	34	64	1156	272
		N = 5	$\sum x_i = 27'7$	$\sum y_i = 126$	176'09	3456	775'1

a) Mean of X: $\bar{x} = \dfrac{26'5}{5} = \boxed{5.54}$　　　　　　Mean of Y: $\bar{y} = \dfrac{27}{5} = \boxed{25.2}$

c) Variance of X: $\sigma_x^2 = \dfrac{176'09}{5} - 5.54^2 = 4.526 \Rightarrow \sigma_x = +\sqrt{4'526} = \boxed{2.1275}$

d) Variance of Y: $\sigma_y^2 = \dfrac{3456}{5} - 25.2^2 = 56.16 \Rightarrow \sigma_y = +\sqrt{56'16} = \boxed{7.4939}$

e) Covariance: $\sigma_{XY} = \dfrac{775'1}{5} - 5.54 \cdot 25.2 = 155.02 - 139.608 = \boxed{15.412}$

Example: Marks in Maths and Physics obtained by 40 students are:

X = Maths marks	3　4　5　6　6　7　7　8　10
Y = Physics marks	2　5　5　6　7　6　7　9　10
Number of students	4　6　12　4　5　4　2　1　2

Calculate means, variances, standard deviations and covariance.

Solution: Do it yourself!!!

x_i	y_i	f_i	$x_i \cdot f_i$	$x_i^2 \cdot f_i$	$y_i \cdot f_i$	$y_i^2 \cdot f_i$	$x_i.y_i.f_i$
3	2	4					
4	5	6					
5	5	12					
6	6	4					
6	7	5					
7	6	4					
7	7	2					
8	9	1					
10	10	2					
		N=40					

8. Pearson linear correlation coefficient

When scatter plots were presented, we observed there is a linear correlation between the variables. Now, it is interesting to quantify this correlation. This is the objective of Pearson linear correlation coefficient, which is defined by the following equation:

Pearson linear correlation coefficient: 　　　$r = \dfrac{S_{xy}}{S_x S_y}$　　　$-1 \le r \le 1$

Sign of indicate the type of correlation between the variables, X and Y:

> ➤ If r > 0, the correlation is **direct**
> ➤ If r < 0, the correlation is **inverse**
> ➤ If r > 0, there is **no correlation**

So:

- If $r = (-1) \rightarrow$ there is a perfect negative linear correlation between X and Y (Figure 1).

- If $r = 1 \rightarrow$ there is a perfect positive linear correlation between X and Y (Figure 2).

- If $r = 0 \rightarrow$ there is not any correlation between X and Y (Figure 3).

Obviously, *r* will have values that will be closer to 1 or (-1) as correlation between variables is bigger.

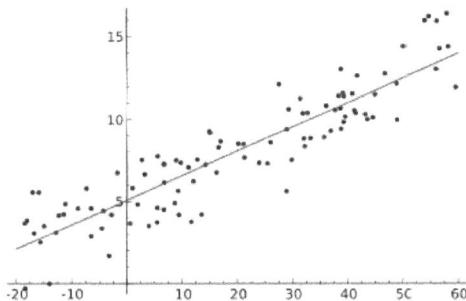   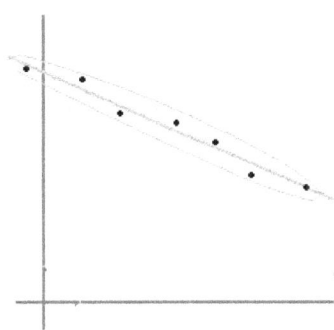   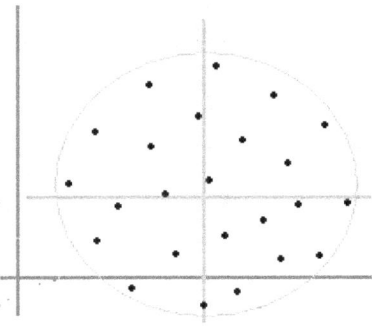

Figure 1   *Figure 2*   *Figure 3*

If between two variables are strongly correlated, the scatterplot condenses around a line. *x* is the independent variable and *y* the dependent variable. Now, the problem is finding the equation of the line that best fits to the scatterplot.

To obtain the equation of the straight line that best fits there are several methods, but the *least squares* method is the most widely used.

Regression straight line of "y on x"

$$y - \overline{y} = \frac{S_{xy}}{S_x^2}(x - \overline{x})$$

Regression straight line of "x on y"

$$x - \overline{x} = \frac{S_{xy}}{S_y^2}(y - \overline{y})$$

With these equations, we can *estimate* values of a variable from known values of the other one.

How good is the correlation?

Remember the closer is |*r*| to 1, the closer are the experimental points to the regression line.

39. A group of 10 friends was brought to a test. They noted the number of hours they spent studying the week before the exam and the mark obtained in the test. The information is contained in the following table:

Hours	21	15	10	15	20	30	18	20	25	16
Mark	9	7	5	2	7	8	8	6	5	4

Represent data as a scatterplot and indicate which of these values you think is the most likely to be the correlation coefficient: 0.92; –0.44; –0.92; 0.44.

40. A survey has been conducted by asking about the number of people living in the family home and the number of rooms of the house. The following table lists the information obtained:

People	3	5	4	6	5	4
Rooms	2	3	4	4	3	3

Calculate covariance and correlation coefficient. How is correlation between both variables?

41. It has been studied, in several printer models, what is the cost per page (in cents) in black and white (B) and in colour (C). The following table gives the first six pairs of data:

X: B	8	11	17	21	14	10
Y: C	33	49	95	106	58	53

a) Find out the regression straight line of Y on X.

b) What is the cost of printing one colour page in a printer model in which a black and white page had a price of 12 cents? Is estimation trustable?

(It is known that r = 0,97).

Review exercises

1. Indicate, for each of the five cases proposed:
- What the population is.
- What the variable is.
- Variable type: qualitative or quantitative; quantitative discrete or continuous.

a) Birth weight of babies.
b) Favourite professions for students.
c) Number of pets is houses.
d) Weekly time spent reading by ESO students in Spain.
e) Number of yellow cards in the last football season.

2. The percentage of vehicles registered during the month of October 2006 is outlined in this table (data are approximate):

Type of vehicle	Percentage
Cars	69
Lorries	17
Motorbikes	
Busses	0.15
Others	1.45

a) Find the percentage of motorcycles registered.
b) Calculate what was the total number of vehicles enrolled, knowing that exactly 279 buses were registered.
c) The set of vehicles registered, is population or sample?
d) Tell what type of variable it is.

3. In a given region, a study on fatal accidents produced at work according to the sector of activity has been done. Here are the results:

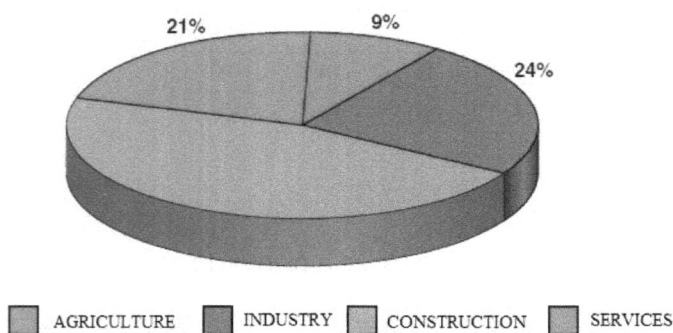

AGRICULTURE   INDUSTRY   CONSTRUCTION   SERVICES

a) What is the percentage of fatal accidents in the construction sector?

b) If there were 135 fatal accidents in agriculture, what was the total number of fatal accidents in the region?

c) How many fatalities occurred in each of the sectors?

4. Figure 1 shows evolution of unemployed ratio in Spain from 1995 to 2000 year. Figure 2 shows a table for the same data, in which only difference is scale for Y-axis.

a) Do both of them give the same sensation?

b) Which one do you think a government would choose to give unemployment data and which one would be chosen by opposition party?

Tables and graphs elaboration

5. When being asked about the number of books they have read for last month, 3rd ESO students answers were the following:

2 1 3 1 1 5 1 2 4 3
1 0 2 4 1 0 2 1 2 1
3 2 2 1 2 3 1 2 0 2

a) Build a frequency table.

b) Draw the corresponding bar diagram.

6. Chosen colour by Spanish people when buying a car is shown in the following table. Draw the corresponding pie chart.

Colour	Percentage
Argent grey	36%
Black	22%
Blue	18%
Red	10%
White	8%
Green	4%
Others	2%

Statistical parameters

7. We have investigated, in different shops, price of a printer, and we have obtained the following results:

146 - 150 - 141 - 143 - 139 - 144 - 133 - 153

    a) Calculate the mean price.   b) What is the median?

    c) Calculate the mean deviation and the range.   d) Calculate the standard deviation.

8. In family A, father's salary is 950 €, and mother's one is 1600 €. In family B, the father earns 1800 €/month, and the mother, 750 €.

    a) What is the mean salary of each family?

    b) In what family is there more dispersion? What is the range for each family?

9. Counting the number of misprints in a book, Peter has obtained the following data. Calculate the
a) mean, b) standard deviation and c) the mode.

Number of misprints	0	1	2	3	4	5
Number of pages	50	40	16	9	3	2

10. In a road speed control, they have obtained the following data:

Speed (km/h)	Number of cars
60 – 70	5
70 – 80	15
80 – 90	27
90 – 100	38
100 – 110	23
110 – 120	17

a) Build a table containing class marks and frequencies.
   (*Clue: class mark of interval 50-60 is 55*).

b) Calculate the mode and the standard deviation.

c) What percentage has speed higher than 90 km/h?

11. Teresa and Rosa are basketball players. Their punctuations for a week training have been:

Teresa	16	25	20	24	22	29	18
Rosa	23	24	22	25	21	20	19

    a) Calculate the mean punctuation for each one.

    b) Calculate their standard deviations and variation coefficients. Who is more regular?

12. 40 people were asked about the number of people living at home:

        4 5 3 6 3 5 4 6 3 2   2 4 6 3 5 3 4 5 3 6

        4 5 7 4 6 2 3 4 4 3   4 4 5 3 2 6 3 7 4 3

    a) Build a frequency table and draw the corresponding diagram.

    b) Calculate the mean, median, mode and standard deviation.

13. When measuring the birth weight of a certain animal species, we have obtained the following data. Represent these data in the appropriated graph. Calculate the mean and standard deviation. What percentage of animal had a birth weight between 5.5 and 6.5 kg? And between 4.5 and 8.5 kg?

Weight (kg)	Number of animals
3.5 – 4.5	1
4.5 – 5.5	8
5.5 – 6.5	28
6.5 – 7.5	26
7.5 – 8.5	16
8.5 – 9.5	1

14. These are the weekly study hours of a group of students:

14	9	9	20	18	12	14	6	14	8
15	10	18	20	2	7	18	8	12	10
20	16	18	15	24	10	12	25	24	17
10	4	8	20	10	12	16	5	4	13

a) Organize these data in a table with the following intervals:

1.5-6.5;   6.5-11.5;   11.5-16.5;   16.5-21.5;   21.5-26.5

b) Build its frequency table and histogram.   c) Calculate the mean and standard deviation.

15. Monthly expenses of a company A have mean 60 000 € and standard deviation of 7500 €. In other company B, mean is 9000 €, and standard deviation is 1500 €. Calculate, by using the variation coefficient, which of them has a higher relative variation.

16. Number of words in each sentence of an economy article is:

17 40 22 25 43 21 17 25 37 12
9 37 32 35 30 21 13 27 41 45
36 40 30 48 45 41 39 39 40 38
28 7 33 35 22 34 23

a) Build a frequency table grouping data into the following intervals:

7 - 13,   14 - 20,   21 - 27,   28 - 34,   35 - 41,   42 - 48

Represent these data in a histogram.

c) Calculate the mean and standard deviation.

17. In my college there are 200 students. They were asked about their favourite pet. See the figure and answer the questions:

a) How many people did answer "dogs"?
b) And "cats"?   **Sol.:** a) 60; b) 50.

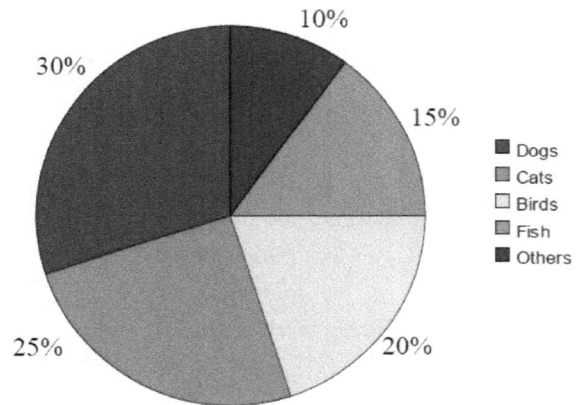

18. Teachers of my college are asked about their favorite film genres. The answers were:
  Comedy: 27%   Action: 18%
  Romance: 14%   Drama: 14%
  Horror: 11%   Foreign: 8%   Fiction: 8%

a) Draw a bar chart and a pie chart with the data.

b) Imagine there are 120 teachers in this high school. How many of them did answer "Romance"?

19. These are the numbers of cars that the different families of a street have:

1 2 1 1 0 2 3 1 1 2
3 1 0 2 1 2 1 1 2 1

a) Complete the table with the frequencies: f_i, h_i, F_i and H_i.

b) Represent the data using a pie chart and a bar graph.

c) Find the mean, median and mode of the data.

20. Blood group (A, B, AB, 0) has been studied in 200 people, as well as their Rh. Some of the results are in the following table.

	A group	B group	AB group	0 group	Total
Rh+	74		6	70	162
Rh-		3	1		
Total				86	200

a) Complete the table.
b) What percentage of population has group B with Rh+?
c) What is the percentage of population that has Rh-?

d) Between people having group Rh+, what percentage do have group A?

21. These three distributions have the same mean, what is it? Their standard deviations are 3.8, 1.3 and 2.9. Join each distribution with its standard deviation. (Do it by observing the graphs, without calculating).

a)   b)   c)

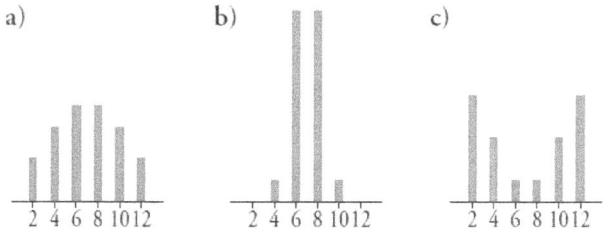

22. A test with 25 questions has been answered by 120 students. Following table shows the results:

Number of correct answers	Percentage
5	10%
15	45%
20	25%
25	

a) Calculate the number of students that answered correctly all the questions.
b) Calculate the mean of corrects answers of population.
c) Calculate the standard deviation.

23. What will happen to mean and standard deviation if we add a same number to all data? And if we multiply them by a same number? Check it with these data: 3, 5, 6, 3, 4, 2, 3, 6.

24. If two distributions have the same mean and first one has a higher standard deviation, which of them will have a higher deviation coefficient?

25. If two distributions have the same standard deviation and first one has a higher mean, which of them will have a higher deviation coefficient?

26. Mean of marks in 3rd A was 6.2, and in 3rd B, 4. If there are 15 students in 3rd A and 35 in 3rd B, what would the mean be if calculated for both groups together?

27. Complete the following table, knowing it mean is 2.7.

x_i	f_i
1	3
2	…
3	7
4	5

28. My final mark in a subject is calculated from marks of three partial exams: second exam's mark has double value, and third one's value is three times the first one.

a) If my marks have been 5, 6 and 4, What will be my final mark?

b) What would my final mark be if each exam had an importance of 10%, 40% and 50%, respectively?

29. My mark is calculated as the mean of four exams. If in the three first I have a mean of 4.2, what should be my mark at the fourth exam to pass the subject?

30. In a company there are 34 employees and 6 directives. Mean salary of all them is 909 €. What is the mean salary of directives if we know that mean of salaries of the rest of workers is 780 €?

31. A stopwatch was used to find the time that it took a group of students to run 100 m.

Time (seconds)	$10 \leq t < 15$	$15 \leq t < 20$	$20 \leq t < 25$	$25 \leq t < 30$
Frequency	6	16	21	8

Estimate the mean and the median.

32. The distances that students in a year group travelled to school is recorded.

Distance (km)	$0 \leq d < 0.5$	$0.5 \leq d < 1.0$	$1.0 \leq d < 1.5$	$1.5 \leq d < 2.0$
Frequency	30	22	19	8

Estimate the median and the mean.

33. The ages of the people at a youth camp are summarised in the table below.

Age (years)	$6 - 8$	$9 - 11$	$12 - 14$	$15 - 17$
Frequency	8	22	29	5

Estimate the mean age, the median, Q_1 and Q_3.

34. The length of telephone calls from an office was recorded. The results are given in the table below.

Length of call (min)	$0 < t \leq 0.5$	$0.5 < t \leq 1.0$	$1.0 < t \leq 2.0$	$2.0 < t \leq 5.0$
Frequency	8	10	12	4

Estimate the mean and standard deviation using this table.

35. The charges (to the nearest $) made by a jeweller for repair work on jewellery in one week are given in the table below.

Charge ($)	$20 - 29$	$30 - 49$	$50 - 99$	$100 - 149$	$150 - 199$	$200 - 300$
Frequency	10	22	6	2	4	1

Use this table to estimate the mean and standard deviation.

36. a) Calculate the standard deviation of the numbers   3, 4, 5, 6, 7.
   b) Show that the standard deviation of every set of five consecutive integers is the same as the answer to part (a).

37. Ten boys sat a test which was marked out of 50. Their marks were

$$28, 42, 35, 17, 49, 12, 48, 38, 24 \text{ and } 27$$

a) Calculate: the mean and the standard deviation of the marks.
b) Ten girls sat the same test. Their marks had a mean of 30 and a standard deviation of 6.5. Compare the performances of the boys and girls.

38. There are twenty students in class A and twenty students in class B. All the students in class A were given an I.Q. test. Their scores on the test are given below.

100, 104, 106, 107. 109, 110, 113, 114, 116, 117,
118, 119, 119, 121. 124, 125, 127, 127, 130, 134.

a) The mean of their scores is 117. Calculate the standard deviation.

b) Class B takes the same I.Q. test. They obtain a mean of 110 and a standard deviation of 21. Compare the data for class A and class B.

c) Class C has only 5 students. When they take the I.Q. test they all score 105. What is the value of the standard deviation for class C?

Two-dimensional statistics

39. In six secondary schools they have studied marks of their students in Maths and in English and they have obtained the following results:

x: Maths	6,5	5,2	6	6,5	7	6
y: English	7	5	5	6	7,5	5

a) Find out the regression straight line of Y on X.

b) Calculate y(5.5). Is estimation trustable? (It is known that r = 0,87).

40. In a driving school they have studied the number of weeks (X) that students had attended and the number of weeks (Y) it takes them to pass the theory test (since it was aimed at the driving school). Data for six students are:

X:	6	1	4	3	5	8
Y:	6	5	5	6	5	10

Find out the regression straight line of Y on X.

41. Marks of ten students in Maths and Physics are the following:

Maths	7	6	4	5	9	10	3	1	10	6
Physics	8	6	3	6	10	9	1	2	10	5

Represent data as a scatterplot and indicate which of these values you think is the most likely to be the correlation coefficient: 0.23; 0.94; –0.37; –0.94.

42. We have measured power (in kW) and the expense (litres/100 km) of six different car models, obtaining the following results:

Power	81	85	66	85	104	83
Expense	7,5	10,6	8,2	9,2	10,7	8,7

Calculate the covariance and the correlation coefficient.

43. A group of six athletes have had a set of tests for length (L) and height (H) jumps. They have had scores between 0 and 5. The results are the following:

Find out the regression straight line of Y on X.

44. We have studied, for different types of yogurts, the percentage of fat they contain and kilocalories per 100 g. These are the results:

X: fat %	2,2	2	1,9	3,1	3	2
Y: kcal	64	55	58	79	65	52

a) Find out the regression straight line of Y on X.

b) How many kilocalories would you expect a yogurt to have if it has 2.5% of fat?

c) And if it has 10% of fat?

Unit 9.- Probability

1. Introduction to probability.

Humans often think in probabilistic terms, whether we are conscious of it or not. That is, we decide to cross the street when the probability of being run over by a car is sufficiently low, we go fishing at the lakes where the probability of catching something is sufficiently high, and so on. So, even when people are wholly unfamiliar with the mathematical formalization of probability, there is an inclination to frame uncertain future events in such terms.

Deterministic and random experiments

In our life, in Nature, we can have two types of experiences or, in probabilistic terms, two types of experiments can be done: *deterministic* and *random* experiments. A **deterministic** experience or experiment is that one in which you can know the result before performing it.

For example, if today it is Monday, we are *sure* tomorrow it will be Tuesday. Or, if I hold a pencil in my hand and I open my fingers, I am sure, even if I do not do it, pencil will fall, it won't go up. So, deterministic experiences are not in the field of study of probability, because they produce *sure* results.

On the other hand, **random** experiments refer to those experiments whose results cannot be determined before performing them. For example, if I throw a coin on a table, I can have two results, and I cannot determine it if I do not throw it. As we cannot be sure about the result of random experiments, we speak about *probability* of a result to happen.

Multiple choice exam	
What good students see	What you see

Probability is the branch of Mathematics that studies and quantifies the chance of random experiences results.

2. Sample set. Elements and events

Sample set. Elements

Each possible simple result of a random experiment is named *element*. The *sample set* of a random experiment includes all the elements in it.

Example: In a box, there are 3 green balls, 2 red ones and 6 blue. Write its sample set. How many elements does it have?

Solution: Sample set: {G, R, B}. Three elements.

Example: In a bag, there are 10 numbered cards, from 0 to 9. Write its sample set. How many elements does it have?

Solution: Sample set: {0, 1, 2, 3, 4, 5, 6, 7, 8, 9}. Ten elements.

Example: Two dice are thrown on a table. We study the sum of them. Write its sample set. How many elements does it have?

Solution: Sample set: {2, 3, 4, 5, 6, 7, 8, 9, 10, 11, 12}. Eleven elements.

Events

> An *event* is a sub-set of the sample set of a random experiment.

Example: We throw a die on a table. As you already know, its sample set is {1, 2, 3, 4, 5, 6}. Describe the following events:

A = Obtain an even number,　　　　　　　　　B = obtain a multiple of three,
C = obtain a number higher than 3,　　　　　　D = obtain a number equal or higher than 3,
E = obtain a number between 1 and 6, both included,　F = obtain number 8.

Solution:　　A = {2, 4, 6}　　　　B = {3, 6}　　　C = {4, 5, 6}　　　D = {3, 4, 5, 6}
　　　　　　E = {1, 2, 3, 4, 5, 6}　　F = Ø (empty set).

There are several special events we must consider:

> - **Sure event** is that one that is always going to happen. Its probability is 100%.
>
> - **Impossible event** is that one that is never going to happen. Its probability is 0.
>
> - Given an event A, we can define its **complementary event**, denoted by $\overline{A}$, as a new event that includes the elements of sample set not included in A. Notice that A + $\overline{A}$ = sample set.

Example: In a box, there are balls numbered from 1 to 10. Describe the following events:
a) A = Obtain a multiple of 4　　　　　　b) B = Obtain a number from 1 to 10
c) C = Obtain number 20　　　　　　　　d) D = Obtain a divisor of 8
e) $\overline{D}$　　　　　　　　　　　　　　f) How could you define elements in $\overline{D}$?

Solution:　　a) A = {4, 8}　　　　　　　　b) B = {1, 2, 3, 4, 5, 6, 7, 8, 9, 10} (sure event)
　　　　　　c) C = Ø (impossible event).　d) D = {1, 2, 4, 8}
　　　　　　e) E = {3, 5, 6, 7, 9, 10}　　f) Numbers from 1 to 10 not being divisors of 8.

Intersection and union of two events

> - Given two events, A and B, **intersection of A and B**, denoted by $A \cap B$, is a new event containing elements that are included, <u>at the same time</u>, in A **and** B.
>
> - Given two events, A and B, **union of A and B**, denoted by $A \cup B$, is a new event containing elements that are included in A **or** B.

Example: Consider the following events of the experiment in which we take a card from a box containing numbered cards from 1 to 10:

A = Obtain an odd number   B = Obtain a number equal or higher than five.

  a) Describe events A and B.   b) Describe complementary event of A, $\overline{A}$

  c) Describe intersection $A \cap B$ d) Describe union $A \cup B$

Solution: a) A = {1, 3, 5, 7, 9}   B = {5, 6, 7, 8, 9, 10}   b) $\overline{A}$ = {2, 4, 6, 8, 10}

  c) $A \cap B$ = {5, 7, 9}   d) $A \cup B$ = {1, 3, 5, 6, 7, 8, 9, 10}

1. Classify the following experiments into deterministic and random experiments:

  a) Extract a card from a cards game.

  b) Check the weight of a litre of mercury.

  c) Ask a number to a friend.

  d) Throwing three dice and write down their sum.

  e) Subtracting two known numbers.

2. Write all the possible results of throwing two coins on a table.

3. We throw a coin and a die. Write the sample set.

4. Given events A = {1, 2, 3} and B = {1, 3, 5}, calculate their union and intersection.

5. When throwing a die with 8 faces, we consider following events:

  A = {2, 4, 5, 8}   and   B = {1, 2, 3, 7}.

Calculate:

  a) $A \cup B$ b) $A \cap B$ c) $\overline{A \cap B}$ d) $\overline{A} \cup \overline{B}$ e) $\overline{A \cup B}$ f) $\overline{A} \cap \overline{B}$

Could you extract a rule by observing results c) and d)?   And e) and f)?

6. One card is extracted from a Spanish card game (40 cards). Two events are considered:

  A = "Getting an even number"   B = "Getting more than a 5"

  Describe the events: *A, B, A', B', A ∪ B* y *A ∩ B.*

7. One card is extracted from a Spanish card game (40 cards). Two events are considered:

  A = "Getting an even number"   B = "Getting more than a 6"

  Describe the events: *A, B, A', B', A ∪ B* y *A ∩ B.*

8. A die is thrown and the result is observed. Two events are considered:

  A = "Getting a divisor of 6"   B = "Getting more than a 4"

  Describe the events: *A, B, A', B', A ∪ B* y *A ∩ B.*

Exercises

As we have already said, probability field is that one in which there are not sureness when performing an experience. So, we can only speak about the probability of an event to happen. In this part of the unit, we are going to learn to quantify this probability.

Expression to quantity probability was developed by Laplace. That is the reason by which it is known as **Laplace's Law.**

Probability of an event is calculated by counting favourable elements for the event and possible elements of the experiment, and calculating their quotient.

$$P(A) = \frac{Number\ of\ favourable\ elements\ to\ A}{Number\ of\ possible\ elements} \qquad \textit{(Laplace's Law).}$$

Example: In a bag, there are ten numbered balls from 1 to 10. A ball is extracted from the bag. Calculate the following possibilities:

a) Obtaining number 8   b) Obtaining an even number
c) Obtaining a multiple of 5

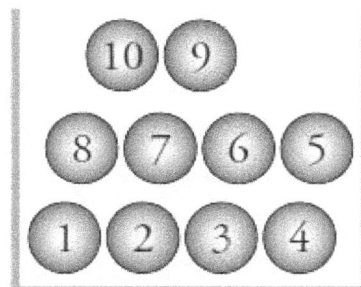

Solution: In all cases, there are 10 possible elements, 10 balls.

a) There is 1 ball with number 8, so, $P(8) = \boxed{\dfrac{1}{10}}$

b) There are 5 even numbers between 1 and 10 (2, 4, 6, 8 and 10), so, $P(even) = \dfrac{5}{10} = \boxed{\dfrac{1}{2}}$

c) There are 2 multiples of 5 between 1 and 10 (5 and 10), so, $P(even) = \dfrac{2}{10} = \boxed{\dfrac{1}{5}}$

---

9. When throwing a die, calculate the probability of obtaining:

a) a multiple of 5   b) a divisor of 2   c) a prime number   d) Number 3

e) a divisor of 6   f) an even number divisor of 4   g) multiple of 7

h) less than 10   i) odd number

10. From a pack of Spanish cards we get a card.   a) What is the probability of getting a horse?

b) And a figure? c) And a "gold"?

11. In a box there are 5 yellow balls and 7 red ones. What is the probability of getting a yellow one? And a green one?

Exercises

12. A card has a black side and white one. We have thrown this card 85 times and we have obtained the white side 43 times. What do you think probability of getting black one is?

13. We throw two dice and sum their punctuations. What is the probability of obtaining:

    a) 3   b) Higher than 10   c) 7   d) 4 or 5.

14. A box contains 4 white balls, 2 red ones and 5 black ones. Calculate the probability to obtain:

    a) a white ball   b) a red ball   c) a white or black ball.

15. In a bag, there are 90 balls, numbered from 1 to 90. We extract one of them.

    a) What is the probability of getting ball number 59?

    b) What is the probability of getting a multiple of 10?

16. In a bag there are balls having two sizes, large (L) and small (S). We extract one of them, write the size (L or S), insert it back to the bag, extract another one, etc. In this way, we observe 84 L balls and 36 S ones. What values would you give to P(L) and P(S)?

17. In a summer school, there are 32 European children, 13 American, 15 African and 23 Asian ones. Representative is randomly chosen. What is the probability of being an European one?

18. Two dice are thrown on a table and sum of punctuations are written. Calculate the probabilities of each sum.

19. In previous experiment:

    a) What is the probability of a sum less than 6?

    b) What is the probability of a sum higher than 6?

    c) And of a sum between 4 and 7, both included?

20. We throw two dice on a table and write down the highest punctuation. If both punctuations are identical, we write down this punctuation. Build table. What is the probability for the highest to be 1? And 2? And 3? And 4? And 5? And 6?

21. One card is extracted from a Spanish card game (40 cards). Two events are considered:

    A = "Getting an even number"   B = "Getting more than a 5"

    Calculate the probabilities: $P[A]$; $P[A']$; $P[B]$; $P[B']$; $P[A \cup B]$; $P[A \cap B]$

22. One card is extracted from a Spanish card game (40 cards). Two events are considered:

    A = "Getting an even number"   B = "Getting more than a 6"

    Calculate the probabilities: $P[A]$; $P[A']$; $P[B]$; $P[B']$; $P[A \cup B]$; $P[A \cap B]$

23. A die is thrown and the result is observed. Two events are considered:

    A = "Getting a divisor of 6"   B = "Getting more than a 4"

    Calculate the probabilities: $P[A]$; $P[A']$; $P[B]$; $P[B']$; $P[A \cup B]$; $P[A \cap B]$

24. In a box, 10 balls numbered from 1 to 10 have been placed. One of them is extracted. Consider these two events: A = "Getting more than 6"   B = "Getting a multiple of 3".

Calculate the probabilities: $P[A]$; $P[A']$; $P[B]$; $P[B']$; $P[A \cup B]$; $P[A \cap B]$

25. A die is thrown and the result is observed. Two events are considered:

A = "Getting an odd number"   B = "Getting less than 6"

Calculate the probabilities: $P[A]$; $P[A']$; $P[B]$; $P[B']$; $P[A \cup B]$; $P[A \cap B]$

26. In a box, 10 balls numbered from 1 to 10 have been placed. One of them is extracted. Consider these two events: A = "Getting an odd number lower than 8"   B = "Getting a multiple of 3".

Calculate the probabilities: $P[A]$; $P[A']$; $P[B]$; $P[B']$; $P[A \cup B]$; $P[A \cap B]$

4. Probability for more complex events

In the following part of the unit, you will learn to calculate the probability of more complex events. You will study situations in which elements can be classified according to two features, as well as experiments composed by two or more actions.

Use of two variable data tables

Two variable data tables are used when describing situations in which each element of the sample or population can be classified according to two features. For example, in a group of people, each person can be classified according to their sex as well as to their eye colour; or citizens of a town can be classified according to the party they vote as well as their financial situation.

An example is presented so that you can understand how to use them.

Example: In a secondary school group there are 31 students and 15 of them are girls. 6 boys and 8 girls use glasses.

a) Complete the table below by using this information.

	Girls	Boys	Totals
Using glasses			
Not using glasses			
Totals			

We randomly choose a person. Calculate the probability of being:

b) A girl   c) A boy who uses glasses   f) Knowing it is a girl, probability of using glasses.

191

Solution:

a) Start writing what you know:

	Girls	Boys	Totals
Using glasses		6	
Not using glasses	8		
Totals	15		31

And now, you will complete it:

- Total number of boys = 31 − 15 = 16.   - Boys not using glasses = 16 − 6 = 10.

- Number of girls using glasses = 15 − 8 = 7.   - Total glasses = 7 + 6 = 13.

- Total "not glasses" = 8 + 10 = 18.

	Girls	Boys	Totals
Using glasses	7	6	13
Not using glasses	8	10	18
Totals	15	16	31

b) Total people: 31; Girls: 15 → $p = \dfrac{15}{31}$

c) Total people: 31; Boys using glasses: 6 → $p = \dfrac{6}{31}$

d) *¡¡¡Notice now, total is not 31 people, but 15 girls!!!*

Total girls: 15; Girls using glasses: 7 → $p = \dfrac{7}{15}$

Exercises

27. In a restaurant there are 28 men and 32 women. 16 men and 20 women have had meat for lunch, and the others, fish. If we randomly choose a person, calculate the probability of the following events: a) Being man   b) Having had fish   c) Being a man and having had fish
d) Knowing he/she has had fish, probability of being a man.

28. In a school, there are 20 boys and 16 girls. Half of the boys and the three fourths of the girls are brown, and the others, blonde. What is the probability that when randomly chosen a person, he/she is a boy or had brown hair?

29. In a town, newspaper A is read by 30% of citizens, and newspaper B is read by 20% of citizens, while 7% of the citizens read both of them. If we randomly choose a citizen, what is the probability that he:

a) Reads some of the newspapers?   b) Does not read any newspaper.

c) Read exactly one newspaper.

30. In a sport centre there are 30 boys and 30 girls. Half of the boys and a third of the girls play tennis. If a person is randomly chosen, calculate the following probabilities:

    a) P(boy)   b) P(does not play tennis)   c) P(boy who does not play tennis)

Use of tree diagrams

Tree diagrams are used when describing situations in which two actions are considered. For example, throwing two coins on a table, or throwing two dice, or extracting two cards from a card game, or choosing two people in a group.

An example is presented so that you can understand how to use them.

Example: A box contains 4 green and 8 blue balls. Two of them are extracted (second one is extracted without having replaced the first one). Calculate the probabilities of getting:
    a) Two blue balls   b) Two balls in the same colour.

<u>Solution:</u>

First of all, we must draw a tree diagram.

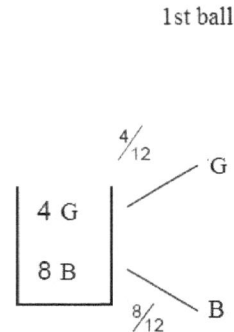

1st ball

- <u>First ball:</u> When we are going to extract the first ball, there are 12 balls in the box. We can extract a green one ($p = \dfrac{4}{12}$) or a blue one ($p = \dfrac{8}{12}$). Write the probabilities on their *branch*.

- <u>Second ball:</u> Notice now there are 11 balls in the box.

    If 1st ball was green, the box contains 3 G and 8 B, so probabilities are now $p(G) = \dfrac{3}{11}$ and $p(B) = \dfrac{8}{11}$.

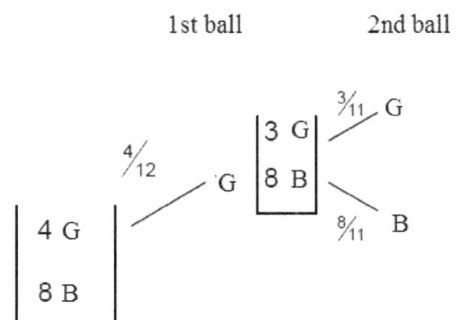

1st ball   2nd ball

If 2nd ball was blue, the box contains 4 G and 7 B, so probabilities are now $p(G) = \dfrac{4}{11}$ and $p(B) = \dfrac{7}{11}$.

1st ball   2nd ball

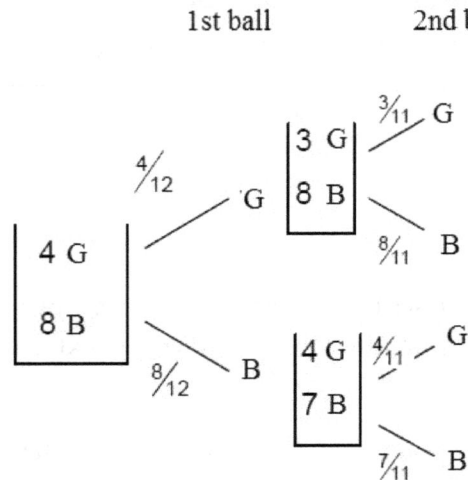

IMPORTANT: The sum of the probabilities on the branches emerging from a common point is **1**.

Check it: $\dfrac{4}{12} + \dfrac{8}{12} = 1$ $\dfrac{3}{11} + \dfrac{8}{11} = 1$ $\dfrac{4}{11} + \dfrac{7}{11} = 1$

Now, you can calculate the asked probabilities:

You must multiply the probabilities of the branches that form a correct combination.

---

a) Getting two blue balls: There are four possible combinations, but only the last one fulfils our

condition (BB)   → $p(BB) = \dfrac{8}{12} \cdot \dfrac{7}{11} = \dfrac{2}{3} \cdot \dfrac{7}{11} = \boxed{\dfrac{14}{33}}$

b) Getting two balls in the same colour: There are two combinations fulfilling our condition (BB, GG):

$p(BB, GG) = \dfrac{8}{12} \cdot \dfrac{7}{11} + \dfrac{4}{12} \cdot \dfrac{3}{11} = \dfrac{14}{33} + \dfrac{3}{33} = \boxed{\dfrac{17}{33}}$

When replacing the first ball before extracting the second one

---

Try to recalculate the probabilities of the previous example, but supposing that we replace the first ball before extracting the second one.

IMPORTANT: The only difference is that in the second extraction, the composition of the box is the same than before the first extraction.

---

31. Two dice are thrown on a table. Calculate the probability of getting:

a) Two "ones".   b) Two numbers being different from 1.

32. In a bad there are 5 balls numbered from 1 to 5. If we extract two of them, replacing the first one before extracting the second one, calculate the probability of getting:

a) Two even numbers   b) An odd and an even number.

33. If two cards are extracted from a Spanish card game (40 cards), calculate the probability of getting: a) Two kings   b) Two figures.

34. A box contains 6 white and 14 black balls. Two of them are extracted (second one is extracted without having replaced the first one). Calculate the probabilities of getting:

a) Two white balls   b) Two balls in different colours.

35. We write the letters of the word PENCIL in six cards and put them in a bag. We extract two of them. Calculate the probabilities of getting:

a) Two vowels   b) A vowel and a consonant.

36. Malik travels to Kingston from Montego Bay on the early bus. The probability that he arrives late is 1/10. He catches the bus on two consecutive days. What is the probability that he arrives:

a) on time on both days,   b) on time on at least one day,   c) late on both days.

Conditional probability

The probability of an event occurring given that another event has already occurred is called a **conditional probability**.

Recall that when two events, A and B, are _dependent_, the probability of both occurring is

$$P(A \text{ and } B) = P(A) \times P(B \text{ given } A)$$

37. In an exam, two reasoning problems, 1 and 2, are asked. 35% students solved problem 1 and 15% students solved both the problems. How many students who solved the first problem will also solve the second one?

38. Out of 50 people surveyed in a study, 35 smoke in which there are 20 males. What is the probability the if the person surveyed is a smoker then he is a male?

39. The probability of raining on Sunday is 0.07. If today is Sunday then find the probability of rain today.

40. In a school the third language has to be chosen between Hindi and French. If a student has taken French then what is the probability that he will take Hindi, if the probability of taking Hindi is 0.34?

41. The probability that it is Friday and that a student is absent is 0.03. Since there are 5 school days in a week, the probability that it is Friday is 0.2. What is the probability that a student is absent given that today is Friday?

42. At Kennedy Middle School, the probability that a student takes Technology and Spanish is 0.087. The probability that a student takes Technology is 0.68. What is the probability that a student takes Spanish given that the student is taking Technology?

Review exercises

1. Classify the following experiments into deterministic and random:
  a) Extract a card from a pack of cards.
  b) Measuring the hypotenuse of a right triangle whose legs are 3 cm and 4 cm.
  c) Pushing "ON" in a lamp.
  d) Measuring the height of a classroom.
  e) Throwing a stone and measuring gravity constant.
  f) Finding out the result of a football match before being played.

2. Write the sample set of the following random experiments:
  a) Extract a card from a pack of Spanish cards.
  b) Extracting a ball from a box containing 5 red, 3 blue and 2 green ones.
  c) Throwing two dice and subtracting the results.
  d) Throwing two dice and multiplying the results.

3. We throw a dodecahedron shaped die with faces numbered from 1 to 12.
  a) What is the sample set?
  b) Write the events: A = less than 5;   B = more than 4;
                           C = even number;   D = Not a multiple of 3.

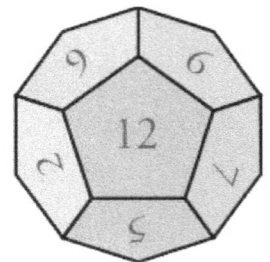

4. We write each letter of the word GAME in a different card, insert them into a bag and extract one.
  a) Write the sample set.   b) Describe the event "obtaining a vowel".
  c) If the word was PROBABILITY, how would you answer?

5. Consider the experiment in which we randomly get a domino token.
  a) Describe the event A = Choosing a token whose numbers' sum is 6.
  b) Describe the event B = Choosing a token whose numbers' product is 12.
  c) Calculate the union and intersection of A and B.

6. In a box there are 15 numbered balls from 1 to 15 and we extract one of them. Write the elements of the following events:
  a) Multiple of 3   b) Multiple of 2   c) Higher than 4
  d) higher than 3 and lower than 8   e) Odd number.

7. When throwing a die with 6 faces, A = {2, 4} and B = {1, 2, 3}. Calculate:
  a) A∩B   b) A∪B   c) The complementary of events A, B, A∩B and A∪B.

8. We throw a die with 6 faces and consider events A = {1, 3, 5, 6}, B = {1, 2, 4, 5} and C = {3, 4}.
Calculate: a) $\overline{A}$ b) $\overline{B}$ c) $\overline{C}$ d) A∪B   e) A∩ B   f) A∪B
          g) $\overline{A \cup B}$ h) $\overline{A} \cap \overline{B}$ i) $\overline{A} \cup \overline{B}$

9. Calculate the probability of obtaining a 5 when throwing each of the following dice:

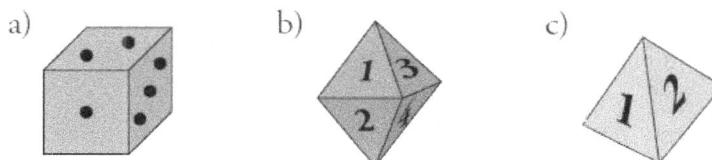

10. In a bag there are 6 red bolls, 4 blue, 7 green, 2 yellow and 1 black one. We randomly extract one of them. Calculate the probability of obtaining one being:
  a) Blue   b) Not black   c) Red or green   d) Nor yellow neither black.

11. In a test exam there are 80 units, from which you choose one randomly. A person has studied 60 of them. Calculate the probability of a) passing and b) failing the exam.

12. Calculate the probabilities of the following events when throwing a correct die:
  a) Multiple of 3   b) Multiple of 2   c) Higher than 1   d) Lower than 5
  e) Lower than 1

13. In a school, there are 990 students, being 510 of them, girls. If we randomly choose a student, what is the probability of choosing a boy?

14. In a high school, students are distributed as indicated in the table:

1st ESO	2nd ESO	3rd ESO	4th ESO	1st BACH	2nd BACH
210	250	260	220	140	120

If we randomly choose a student, what is the probability that he/she studies:
a) 3rd ESO   b) ESO   c) Bachiller

15. In a book having 120 pages, we have counted the number of misprints in each page. Results are in the following table. When randomly chosen a page of this book:

Number of errors	Number of pages
0	58
1	42
2	16
3	3
4	1

    a) What is the probability of choosing a page without misprints?
    b) What is the probability of choosing a page having exactly 2 misprints?
    c) What is the probability of choosing a page having misprint/s? And having more than 3?

16. We have a card having a white and a black faces. We throw it 3 times. What is the probability of obtaining the black face a) exactly once?   b) twice and b) three times?

17. We randomly extract a domino token. Calculate the probability of getting:
    a) A sum of punctuations being lower than 4.
    b) A sum of punctuations being multiple of 3.
    c) A "Double" token (having both identical punctuations).

18. We throw two dice: Calculate the probability of obtaining
    a) A product of punctuations of 5.
    b) A product of punctuations of 6.
    c) A product of punctuations of 4.   *(Clue: build a table with possible results).*

19. We throw 100 times a tetrahedron shaped die (faces: 1, 2, 3 and 4), writing the number of the hidden face. We have obtained the following results:

Face	1	2	3	4
Frequency	28	22	30	20

Calculate the probabilities of the following events:
    a) Multiple of 3   b) Multiple of 2   c) Higher than 1   d) Lower than 1.

20. Probability of an event is 0.2. What is the probability of its complementary?

21. If in a die, $P(1) = P(2) = P(3) = 0,14$ y $P(4) = P(5) = P(6) = x$, what s the value of x?

22. In a non-correct die, probabilities of each face are the following:

Face	1	2	3	4	5	6
P(i)	0.1	0.1	0.1	a	b	0.4

Knowing that $P(4) = 2 \cdot P(5)$, what are the values of *a* and *b*?

23. We randomly choose a number from 1 to 30. Consider the events A = "Obtaining an even number being lees or equal to 14", B = "Obtaining a multiple of 3 less or equal to 10" and C = "Obtaining a multiple of 10". Calculate the probabilities of:

a) A∪B   b) A∪C   c) A∪$\overline{B}$ d) C∪B   e) B∩Cf) $\overline{A}$ ∩ B

24. Consider a game in which you throw two dice and you win if you obtain a sum of punctuations of 11 or 7. What is your probability of winning?

25. Numbers in a bingo's roulette are from 1 to 36. Calculate the following probabilities:

a) Getting number 9.

b) Getting an even number.

c) Getting a number finishing in a 3.

d) Getting a multiple of 4

e) Getting a number having two identical digits

f) Getting a number starting with a 4.

26. Luis and Juan must clean their room. Luis puts in a bag 3 red balls, 2 green and 1 blue, and purposes to his brother to get one of them. If it is red, Juan will clean, and if it is blue, Luis will clean.
  a) Is this a fair system?
  b) Juan does not accept this system and he purposes another one to his brother: if he gets a red ball, he will clean, and if it is blue or green, Luis will clean. Is this a fair system?

27. Victoria calls for her friends, Kina and Freya. The probability that Kina is not ready to leave is 0.2 and the probability that Freya is not ready is 0.3. Use suitable tree diagrams to find the probability that:

a) both Freya and Kina are ready to leave,

b) one of them is not ready to leave,

c) Kina is not ready to leave on two successive days,

d) Kina is ready to leave on two consecutive days.

28. A bottle contains 20 balls, between black, red and green. We do not know how many there are with each colour, because the bottle is not transparent. We can only see into the bottle in a very little transparent part at the top of the bottle, where there is exactly the space for one ball. We can rotate the bottle down and see this ball. We have moved and rotated the bottle for 1000 times, obtaining the following frequencies:

f(Black) = 461 $\qquad\qquad$ f(Red) = 343 $\qquad\qquad$ f(Green) = 196

We can estimate the number of balls of each colour. We are doing it for you with black ones.

$$f\text{(Black)} = 461 \quad \rightarrow \quad h\text{(Black)} = \frac{461}{1000} = \boxed{0.461}$$

There are 20 balls. Let's suppose there are "b" black balls. Probability of seeing a black ball will be:

$P\text{(Black)} = \dfrac{n}{20}$ $\quad$ By equalizing to the corresponding frequency, $\dfrac{n}{20} = 0.461 \rightarrow n = 20 \cdot 0.461 = 9.22$.

So, we can estimate that there are $\boxed{9 \text{ black balls}}$.

Now, do it for the others.

$$\binom{0}{0}$$

$$\binom{1}{0} \quad \binom{1}{1}$$

$$\binom{2}{0} \quad \binom{2}{1} \quad \binom{2}{2}$$

$$\binom{3}{0} \quad \binom{3}{1} \quad \binom{3}{2} \quad \binom{3}{3}$$

$$\binom{4}{0} \quad \binom{4}{1} \quad \binom{4}{2} \quad \binom{4}{3} \quad \binom{4}{4}$$

$$\binom{5}{0} \quad \binom{5}{1} \quad \binom{5}{2} \quad \binom{5}{3} \quad \binom{5}{4} \quad \binom{5}{5}$$

Unit 10.- Combinatory

1. Introduction to combinatory.

You think you can count, do you? Answer these:

- *How many ways can 8 books be arranged on a bookshelf?*
- *If you have 20 employees and you need to form a committee of 5, how many ways can you do it?*
- *How many possible ways are there to pick the numbers for the lottery?*
- *How many possible 9-digit phone numbers are there?*

Factorial of a number (symbol: !) is the function that produces the product of the number by a series of its descending natural numbers.

$$n! = n \cdot (n-1) \cdot (n-2) \cdot \ldots \cdot 1$$

Examples:
- $4! = 4 \times 3 \times 2 \times 1 = 24$
- $7! = 7 \times 6 \times 5 \times 4 \times 3 \times 2 \times 1 = 5{,}040$
- $1! = 1$

Important: $\boxed{0! = 1}$

3. Combinatorial numbers

$$C_m^n = \binom{m}{n} = \frac{m!}{n! \cdot (m-n)!} \qquad \text{where } m > n$$

Example:

$$C_5^2 = \binom{5}{2} = \frac{5!}{2! \cdot (5-2)!} = \frac{5!}{2! \cdot 3!} = \frac{5 \cdot 4 \cdot 3!}{2! \cdot 3!} = \frac{5 \cdot 4}{2 \cdot 1} = 10$$

4. Elements of combinatory

In this unit, you will learn to count groups of elements without counting them. But, first of all, you must know the ways in which elements can be grouped: variations, permutations and combinations.

Ordered groups: two groups with the same elements, in a different order, are different groups. For example, ABC and ACB are different groups. Ordered groups can be **variations** and **permutations**.

Not ordered groups: If two groups have the same elements, they are the same group, not depending on the order of their elements. For example, ABC and ACB are the same group. Not ordered groups are **combinations.**

Repeated or no repeated elements? This is another aspect we will have to take into account.

Variations and *permutations* are groups in which *order is important*. The difference between them is the number of elements of the set you get to build the subset.

While in *variations* you get only *some* of the elements of the set, in *permutations*, you get *all* the elements of the set.

In this part of the unit, you will work on variations.

Variations with repetition

> **Variations:** Ordered group in which we only take *n* elements among *m* elements.
>
> Number of variations of *n* elements taken among *m* elements. Elements can be repeated:
>
> $$VR_{m,n} = m^n$$

Example: How many words of 2 letters can you make with 4 letters a, b, c, d? You can repeat them.

Solution:

By direct counting:
{aa} {ab} {ac} {ad}
{ba} {bb} {bc} {bd}
{ca} {cb} {cc} {cd}
{da} {db} {dc} {dd}   There are $\boxed{16}$ total variations on the set.

By using the expression: m=4; n=2. The order of the elements does not matter, so, it is a variation. Notice that the elements can be repeated. $VR_{4,2} = 4^2 = \boxed{16}$.

Variations without repetition

> **Variations:** Ordered group in which we only take *n* elements among *m* elements.
>
> Number of variations of *n* elements taken among *m* elements (without repeating them):
>
> $$V_{m,n} = \overbrace{m \cdot (m-1) \cdot (m-2) \cdot (m-3) \cdot \dots\dots}^{n\ factors}$$

Example: How many different three-digit numbers can be formed with the digits 1, 2, 3, 4, 5?

Solution: m=5; n=3. The order of the elements does matter, so, it is a variation. Notice that the elements cannot be repeated. $V_{5,3} = 5 \cdot 4 \cdot 3 = \boxed{60}$.

Example: 10 candidates have been presented with awards for their novels in a literary contest. The honour roll (list of candidates) is formed by the winner, finalist and runners-up. How many different honour rolls can be formed?

Solution: m=10; n=3. The order of the elements does matter, so, it is a variation. Notice that people cannot be repeated. $V_{10,3} = 10 \cdot 9 \cdot 8 = \boxed{720}$.

<table>
<tr>
<td rowspan="6" style="writing-mode: vertical-lr;">Exercises</td>
<td>

1. How many different flags of three horizontal stripes of different colours can be made up from seven different colours?

2. How many different four-digit numbers can be formed with the digits 2, 4, 6, 8 and 9?

3. With the letters of the word PENCIL, how many words, with or without meaning, can be formed with 4 letters? They can be repeated.

4. In a handball tournament there are 8 teams and only 3 trophies, haw many different podiums can be formed?

5. In a firm, the headmaster and the representative may not be covered by the same person, calculate how many possibilities there are to cover these positions if there are 22 employees.

6. The current system of car plates combine 4 digits with 3 letters, which are chosen from 10 numbers and 26 letters. How many different plates are possible?

</td>
</tr>
</table>

6. Permutations

In permutations, as well as in variations, order does matter. The difference between Permutations and Variations is the number of elements you take. In Permutations, we have *m* elements and we take **all of them**.

So, expressions to calculate the number of permutations can be deduced from those to calculate variations.

Permutations are always without repetition. Notice that a permutation is a variation in which $n = m$.

$$P_m = V_{m,m} = \overbrace{m \cdot (m-1) \cdot (m-2) \cdot (m-3) \cdot \ldots}^{m \ factors} = m \cdot (m-1) \cdot (m-2) \cdot \ldots \cdot 1 = m!$$

Permutations: Ordered group in which we take **all** the element.

Number of permutations of *m* elements:

$$P_m = m!$$

Example: How many ways can eight people sit in a row of seats?

Solution: The order of the elements does matter and all the elements are used, so, it is a permutation.
$$m = 8 \quad \rightarrow \quad P_8 = 8! = \boxed{40320}.$$

7. The 13 students in a group of 2^{nc} year want to make a picture all together in a row. In this photo should not appear or two girls neither two boys together. Knowing that there are 7 girls, how many different ways they can be placed?

8. Eight cyclists are going along a straight path. How many ways can they be arranged?

9. With all the letters of the word TABLES, how many words, with or without sense, can be formed without repeating letters?

10. In a race in a school, 6 students take part. How many different orders can they arrive?

11. With the letters of the word TUESDAY, how many words, with or without sense, can be formed without repeating letters?

12. With the odd digits, how many different five-digit numbers can be formed?

13. How many ways can 4 CDs be kept into 4 different folders?

14. In how many ways can 11 books be arranged on a shelf using all the books?

15. In how many ways can the letters of the word MONDAY be arranged using all six letters?

7. Combinations

A *combination* is the choice of n elements from a set of m elements, where order does **not** matter.

In this part of the unit, you will work on variations.

Combinations without repetition

Combinations: Not ordered group in which we only take n elements among m elements.

Number of combinations of n elements taken among m elements. Elements cannot be repeated:

$$C_{m,n} = \frac{V_{m,n}}{P_n} = \frac{\overbrace{m \cdot (m-1) \cdot (m-2) \cdot (m-3) \cdot \ldots\ldots}^{n\ factors}}{n!}$$

Example: A student council committee of three students has to be selected out of a class or 35 students. How many different committees can be formed?

Solution: The order of the elements does not matter; the elements cannot be repeated.

$$C_{35,3} = \frac{V_{35,3}}{P_3} = \frac{35 \cdot 34 \cdot 33}{3 \cdot 2 \cdot 1} = \boxed{6545}.$$

Exercises

16. Jose has 9 friends that he wants to invite to dinner but he can only invite six of them at one time. Out of the nine friends many different groups can he invite?

17. To make a paddle team, 4 players and a coach are needed, which must be chosen among a group of 10 players and 3 coaches. How many different teams can be formed?

18. A family of 6 people are the winners of a trip for two people. How many ways they share the journey?

19. For breakfast, Mario has to choose among four different biscuit among the 12 different types he has. How many possible choices can he do?

20. How many 5 player basketball teams can be made up with 11 players?

21. Belén needs to choose 4 people among 20 candidates to make her working team. How many ways can she make the selection?

Review exercises

1. How many even 3-digit positive integers can be written using the digits 1, 3, 4, 5, and 6?

2. In how many different ways can a 8-question true-false test be answered:

   a) if every question must be answered?

   b) if it is all right to leave questions unanswered (i.e. answer true, false, or leave blank)?

3. In how many ways can you select 4 cards, one after the other, from a 52-card deck:

   a)   if the cards are returned to the deck after being selected?

   b)   if the cards are not returned to the deck after being selected?

4. How many 7-digit phone numbers can be created if the first digit cannot be 0 or 1, the second must be a 5, and the third must be a 3 or 4?

5. Four cards numbered 1 through 4 are shuffled and 3 different cards are chosen one at a time without replacement. Make a tree diagram showing the various possible outcomes.

6. How many license plates of three symbols (letters and digits) can be made using at least 2 letters each?

7. In how many ways can 11 books be arranged on a shelf:

    a) using all the books?

    b) using 4 of the books?

8. A sample of 4 IPods taken from a batch of 100 IPods is to be inspected. How many different samples can be selected?

9. In a group of 10 people, each person shakes hands with everyone else once. How many handshakes are there?

10. In how many ways can 4 people be seated in a row of 12 chairs?

11. In how many ways can a committee of 7 be chosen from 9 girls and 8 boys if

    a) all are equally eligible?

    b) the committee must include 4 girls and 3 boys?

12. How many different 6-character license plates can be made with the first three characters as letters and the last three as digits

    a) with repeats

    b) without repeats

13. In how many ways can the letters in the word

    a) PERU be arranged?

    b) CANADA be arranged?

14. How many 7-card hands can be dealt having

    a) exactly 3 aces?

    b) exactly 3 of a kind?

15. How many even 3-digit positive integers can be written using the digits 1, 2, 4, 7, and 8?

16. In how many ways can up to 4 students be selected from 6 girls and 7 boys if each selection must have an equal number of girls and boys?

17. Simplify: $_{n+2}P_n$

18. How many 13-card hands having exactly 10 from any suit be dealt?

19. A high school baseball coach must decide on the batting order for a team of 9 players.

    a) How many different batting orders are possible?

    b) How many batting orders are possible if the pitcher bats last?

    c) How many different batting orders are possible if the pitcher bats last and the team's best hitter bats 3rd?

20. How many 3 digit numbers contain no 7s?

21. How many 3-difit numbers contain at least one seven?

22. Three couples go to the movies and sit together in a row of 6 seats. In how many ways can these people arrange themselves if each couple sits together?

23. Three door prizes are to be given to 3 lucky people in a crowd of 100.

    a) If the three prizes are identical, in how many ways can this be done?

    b) If the three prizes are different (1st, 2nd, and 3rd), in how many ways can this be done?

24. A track coach must choose a 4-person 400 m relay team and a 4-person 800 m relay team from a squad of 7 sprinters, and of whom can run on either team. If the fastest runner sprints last in both races, in how many ways can the coach form the two teams if each of the 6 remaining sprinters runs only once and each different order is counted as a different team?

25. If you have five signal flags and can send messages by hoisting one or more of them on a flag pole, how many messages can you send?

26. Baskin Robbins has 31 flavours. How many two scoop cones can be made if order is important and flavours cannot be repeated?

35. How many three scoop cones are possible if flavours can be repeated?

36. A town council consists of 8 members including the mayor.

    a) How many different committees of 4 can be chosen from this council?

    b) How many of these committees include the mayor?

    c) How many do not include the mayor?

    d) Relate the above to Pascal's Triangle.

37. How many ways can 8 jackets of different styles be hung

    a) a straight bar?

    b) on a circular rack?

38. There are three roads from town A to B, 5 roads from town B to town C, and 4 roads from town C to town D. How many ways are there to go from A to D via B and C?

39. If you have a $1 bill, a $5 bill, a $10 bill, and a $20 bill, how many different sums of money can you make using one or more of these bills?

40. Meaty Pizza Dude offers the following toppings: pepperoni, sausage, beef, ham, and anchovies. How many different pizzas can be made?

41. A railway has 30 stations. On each ticket, the departure station and the destination station are printed. How many different tickets are possible?